The
Lobster
Chronicles

*Also by Linda Greenlaw
in Large Print:*

The Hungry Ocean

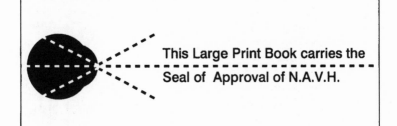

This Large Print Book carries the
Seal of Approval of N.A.V.H.

The Lobster Chronicles

LIFE ON A VERY SMALL ISLAND

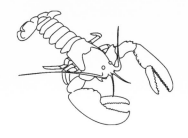

LINDA GREENLAW

Thorndike Press • Waterville, Maine

Published in 2002 by arrangement with Hyperion, an imprint of Buena Vista, Inc.

Thorndike Press Large Print Americana Series.

The tree indicium is a trademark of Thorndike Press.

The text of this Large Print edition is unabridged.
Other aspects of the book may vary from the original edition.

Set in 16 pt. Plantin by Christina S. Huff.

Printed in the United States on permanent paper.

Library of Congress Cataloging-in-Publication Data

Greenlaw, Linda, 1960–
    The lobster chronicles : life on a very small island / Linda
    Greenlaw.
      p. cm.
    ISBN 0-7862-4824-6 (lg. print : hc : alk. paper)
    1. Lobster fisheries — Maine — Isle au Haut. 2. Greenlaw,
    Linda, 1960–   3. Isle au Haut (Me.)   4. Large type books.
    I. Title.
    SH380.2.U6 G74 2002b
    639′.54′0974153—dc21   2002032098

This book is dedicated to
my mother and friend,

Martha Louise Greenlaw.

*"The winters drive you crazy, and the
fishing's hard and slow,
You're a damned fool if you stay,
but there's no better place to go . . ."*

—GORDON BOK,
FROM HIS SONG
"THE HILLS OF ISLE AU HAUT"

CONTENTS

NOTE FROM THE AUTHOR

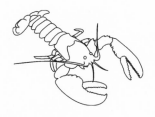

I am presently in the midst of my fifth lobster-fishing season, which ranks me among the amateurs, as opposed to the professionals who have been thinking like lobsters for fifty years or better. I've learned enough to give the appearance of knowing what I'm doing, as I go through the motions of setting and hauling traps, but do not yet consider what I'm doing fishing. The season chronicled in this book is *my* season, filled with personal observations, reflections, and opinions that represent only Linda Greenlaw. If each of the twelve active commercial lobstermen on the Island were to write a book about a particular lobster season, there would be a dozen very different accounts. (The only thing they would all share, though, would be much talk about the difficulties in making ends meet. In order to reflect that, I have omitted mentioning the improvement in my own financial status that followed the publication of my first and only other book, an event that occurred several seasons into my lobstering career.) And although I have used real names in most instances throughout, I have taken the liberty of composite characters in a few cases, and have changed a couple of names and dates and details in others, not to protect the innocent, but in hope of being allowed to remain on the Island after publication. I have also played with the chronology from time to time. While this book tells of a typical season of lobstering, the season it chronicles would prove to be unlike any I would ever experience. It would start normally enough, but would become a season when I would reexamine all I ever thought I knew about myself, life, and lobsters.

LOBSTERS

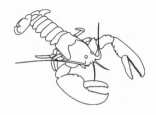

In terms of status, the lobster has come a long way. *Homarus americanus,* or the Maine lobster, ascended from humble fare to fodder fit for royal banquets in a relatively short one hundred years, a true success story. Prior to the nineteenth century, only widows, orphans, and servants ate lobster. And in some parts of New England, serving lobster to prison inmates more than once a week was forbidden by law, as doing so was considered cruel and unusual punishment.

Lobsters are *Arthropoda,* the phylum whose membership includes insects and spiders. Although lobsters are highly unsightly, the sweet, salty, sensual delight of a claw dipped into drawn butter more than compensates for the lobster's cockroachlike appearance and the work involved in extracting meat from shell. Yet in spite of prestige and high standing, the fishermen of Isle au Haut still refer to them as "bugs."

Isle au Haut (pronounced I-LA-HOE) is a small inhabited island off the coast of Maine in an area regarded as "the lobster capital of the world," Penobscot Bay. In a lobster fishing community such as Isle au Haut, the calendar year can be best described as a two-season system: the lobster season and the off-season. Because this is true of all fishing communities up and down the coast, and because residents rarely refer to their home by name, Isle au Haut will be referred to throughout this book as simply "the Island."

Friends fear the exploitation of our Island, and worry that any mention of its name will result in increased traffic to our precious and quiet rock. However, many travel articles in magazines and newspapers (not to mention television features) have run over the years, all touting the wonders of various aspects of life and events on Isle au Haut, and all this attention has thankfully failed to transform us into the dreaded Coney Island. So I suppose I should be flattered that my friends think it possible that my readership might do just that. Oh, I admit that years ago, when I read a *Parade* magazine article about the Island's three Quinby children, who the journalist claimed were all geniuses, I briefly feared that every parent on the planet desiring gifted, talented, exceptional offspring might attempt to move here, hoping that this concentration of brains might be the result of something in the air, or the water, rather than of Quinby genes. Happily, nobody came.

Still, as a way of placating my nervous friends, family, and neighbors, I want to make it clear that in addition to the reasons stated above, I am calling Isle au Haut "the Island" because it really is representative of any piece of land surrounded by water that is inhabited by hardworking, independent people, most of whom are lobstermen. If by any chance, in the course of reading this book, you should fall in love with, or become consumed with curiosity about Maine island life, I promise you that visiting Mount Desert Island, Bailey Island, or Monhegan will surely satisfy both lust and curiosity. People there welcome tourism. They have hotels and restaurants. We have nothing.

Well, not exactly nothing. The list of what we *do* have is shorter than that of what we do *not* have, and those of us who choose to live here do so because of the length of both lists.

We have what I believe could be the smallest post office in the country, and a privately owned boat contracted to haul U.S. mail on- and off-Island. We currently have forty-seven full-time residents, half of whom I am related to in one way or another. (Family trees in small-town Maine are often painted in the abstract. The Greenlaws' genealogy is best described in a phrase I have heard others use: "the family wreath.") We have one general store, one church, one light-house, a one-room schoolhouse for grades K through eight, a town hall that seconds as the school's gymnasium, three selectmen, a fish-ermen's co-op, 4,700 rugged acres of which 2,800 belong to Acadia National Park, and 13 miles of bad road. And we have lobsters.

We do not have a Kmart, or any other mart. We have no movie theater, roller rink, arcade, or bowling alley. Residents can't get manicured, pedicured, dry-cleaned, mas-saged, hot-tubbed, facial-ed, permed, tinted, foiled, or indoor tanned. We have neither fine dining nor fast food. There is no Dairy Queen, Jiffy Lube, newspaper stand, or Starbucks. There is no bank, not even an ATM. No cable TV, golf course, movie the-ater, gym, museum, art gallery . . . Well, you get the picture.

Lobster season for most of us on the Island begins in early May and ends around the first of December. Some fishermen extend or shorten on either end, but in general, we have a seven-month fishing season, and five months of off-season. Each lobster season is typical only in that it is different from every preceding span of seven months in which lobsters have been fished. There are trends, patterns, and habits that are observed by every generation, but each individual season has its own quirks, ebbs, and flows of cooperative crustaceans. Still, there seems to be in the fishermen's credo a tendency to be amazed that the lobsters this season are not acting the way they did last season. And each season every fisherman will attempt to think and reason like a lobster in hopes of anticipating their next move. A lobster's brain is smaller and simpler (in relation to its body mass) than that of nearly any other living thing in which some form of brain resides. So some fishermen are better suited for this game than others. I am not ashamed to admit that I am not among the best lobster fishermen on the Island.

Although the individual members are for the most part hardy, the year-round community on the Island is fragile. This winter's population of forty-seven people is down from seventy residents just two years ago. There are multiple threats to the survival of the community, most notably ever-increasing land values, corresponding property taxes, and extremely limited employment opportunities. The Island, for most of us, is more than a home. It is a refuge. What seems to sustain the community as a whole is lobster. Every year-round family is affected by an abundance or scarcity of income generated by hauling and setting lobster traps. Other than the fact that we all live on this rock, our only common bond is lobster. Island fishermen are presently enjoying the presumed tail end of a lobster heyday, a boom that has endured several seasons of tens of thousands of traps fished and yearly predictions by biologists of sure and pending doom. Our own little piece of America hangs on by a thread to the fate of the lobster.

A small community bears a heavy load. Elderly Islanders move to the mainland when isolated life becomes too strenuous. Why do we not care for our old folks? Small-town politics creates rifts and scars so deep that some individuals, in fact entire families, have found reason to seek opportunity off-Island. Some who remain are nearly hermits, reclusive family units, couples, and individuals preferring seclusion. Man-made problems are inherent in Island life. Yet in our minds, all boils down to the lobster.

Lobsters are tangible. Lobsters become the scapegoat, or perhaps it would be more accurate to say that all threats to our ability to catch lobsters become scapegoats. We have no control over Mother Nature, so she is the easiest target. A major storm could wipe us out, boats and gear gone. Disease has been held responsible for catastrophic lobster-kills throughout the fishery's history. Runoff of chemicals and insecticides has devastated stocks in distant grounds quite recently. I moved back to the Island for many reasons, one of which was my desire to make a living fishing for lobster. Upon my return, it became abundantly clear that the greatest hindrance to my happiness and financial welfare would be what all Islanders perceive as the most palpable threat to our livelihoods: the overfishing of our Island's fishing grounds by outsiders. The threat from mainland lobstermen was both real and present, and was increasing exponentially with each new season. It dwarfed any threat Mother Nature had recently made. At the time of my joining the game, it was clear that the situation would culminate in war.

LAUNCHING

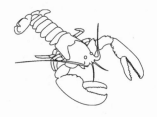

Too warm to fall and hug the ground, yet too cool to rise, the steam hung in the air, a condensed mist as thick as melted pearls. "Pea soup," that's what old-timers would call this wet, raw oyster world of fog. No, I thought, as I made my way down the dock, this was not Carl Sandburg's fog that came in "on little cat feet." This fog had arrived in Stonington, Maine, driving a Mack truck. Although eighteen-wheelers are rarely seen winding through Stonington's narrow streets, fog is a frequent visitor. Stonington is the capital, as it were, of Deer Isle, which is connected to the mainland by a bridge and is, for us Islanders, mainland, too. It is where I store my boat off-season.

Fog is usually associated in books and films with suspense, but there wasn't any mystery surrounding my activity this afternoon. I would soon climb aboard my just-launched lobster boat, the *Mattie Belle*, and head back to the Island as I had every spring since leaving, four years ago, a boat called the *Hannah Boden* and a seventeen-year career of chasing swordfish around the North Atlantic. I slung two garbage bags full of clothes over the rail and onto the deck of my 35-foot Duffy, where they came to rest with two consecutive "plops." Weren't these the same two bags filled with the same stinking clothes that I had reluctantly stevedored off of the *Hannah Boden*? It seemed like just yesterday that I decided to move home to the Island and back in with my parents. Hadn't I desired a simpler life? I had certainly achieved that goal, I thought. But now I wondered about the goal I had set. I had not come ahead very far in the years since I had given up offshore fishing. I now stepped from the dock onto the deck of my most valued possession, a toy compared to the massive *Hannah Boden*, and one that I shared with the bank. Other than two bags of old clothes and some fishing gear, I owned nothing outright. I had not accumulated any "stuff," but realized that being unmaterialistic was a situational virtue, not a heartfelt belief.

I remembered vividly how awkward and sad I felt four years ago as I watched the 100-foot, state-of-the-art *Hannah Boden*, which I had come to refer to as "my boat," leave without me as I stood on the dock in Gloucester. I had chosen to leave the *Hannah Boden* to do my own thing, but the steps from Gloucester to Stonington had been littered with second thoughts. Now fully immersed in Island life after four seasons, I felt as I had when heading offshore from the very first dock, the very first time: apprehension, anticipation, hope, optimism, and the allure of record-breaking catches . . . they seduced the fisherman in me every time! If I did well enough this season, I thought, I would buy a piece of land on the Island. Then I would at last be fully vested in this life that I had so carefully planned.

Other than out at sea, the Island was the only place I had ever felt at home. It made sense. Boats and islands have much in common, and whatever quirks and peculiarities made me fit for, and happy with, a life on the ocean would quite naturally be requisite for me to thrive as an Islander. Although all supplies, U.S. mail, and electrical power via underwater cable come to us from the mainland, there is no bridge connecting the Island to Stonington. In fact, the 7 miles of salt water separating the two reinforce those traits of independence and self-sufficiency that are necessary for all Islanders and fishermen.

I have never been a stranger to hard work. Nonfishermen often express a notion of what they imagine commercial fishing to be, which is ultimately romantic. But the truth is, the fun, adventure, excitement, and beauty that you do experience make up a fairly small percentage of the whole. Admittedly, that tiny percentage leaves such a feeling of awe that it more than compensates for the fatigue, monotony, sweat, and frustration — but nothing comes without work.

Preparing the *Mattie Belle* for the launching had taken me only one day, but it had been a long one. I was underneath the *Mattie Belle* at daylight. The most awkward place to reach with a roller full of bottom paint is under the stern, where it flattens out, aft of the rudder. The bottom of the boat is too close to the ground for me to stand up straight under it, and too far to reach from a kneeling position, so I had resorted to an uncomfortable crouch in order to complete the annual application of antifouling paint. Although the application process is physically painful, the last few swipes with the paint roller obliterated lessons learned and cleaned the slate for a new season.

Antifouling paint is weird stuff. It is not as fluid as paint used on a boat's topside. It's thick, with a sappy consistency of honey, and it dries quickly to a chalky finish that sheds when rubbed. The shedding, along with the paint's ingredients, are what keep that portion of the boat's hull below the waterline free of the plants and organisms that like to attach and grow there.

Barnacles, mussels, hair, weeds, and slime adore the unprotected bottom of a boat, and are the most visible sign of negligence on the part of a boat's captain. It is often said that the relationship of fisherman and boat is akin to a marriage. As in any relationship, lack of maintenance and attention precipitates failure. Most lobstermen do the best they can with what they have to work with, some with racy-looking trophy-bride boats and others with spruced up old gals who plod along willingly enough. Billy Barter fishes one of the oldest and one of very few wooden boats in the area. His boat, the *Islander*, is also the prettiest and most immaculately maintained vessel in the thoroughfare. The same can be said of Billy's wife. Oh, Bernadine is by no means the oldest woman on the Island, but she takes great pride in her appearance. She works at staying trim and fashionable, not easy for someone residing on a rock. Bernadine dresses for any occasion or nonoccasion with a strange combination of grace and pizzazz. When she enters a room, much like the *Islander* slipping up to a dock, Bernadine always catches an appreciative eye.

The correlation between the condition of a fisherman's boat and spouse is complex and interesting. There are examples of slovenliness. I'll leave it at that. As for how this all applies to me? Well, first of all, I'm single and my boat is named for my grandmother Mattie Belle Robinson Greenlaw, who passed away long ago. I prefer to think of the relationship as less than intimate. I do, out of respect for the boat and pride in myself, keep the *Mattie Belle* painted and relatively clean.

When I had finally crept out from under the stern, I had straightened my back and looked up at the sun that I'd been shaded from for the nearly three hours it had taken me to complete the nastiest job in preparation for this year's lobster season. Early in May, the Maine sun is seen more than felt. There is a chill in the air that the sun will not overcome for another month. The weatherman had predicted an approaching warm front, which is what would cause the fog in the afternoon. Things were just sort of damp in general at Billings' Diesel.

I had watched four men untie and pull a tarpaulin off a sailboat as I cleaned blue paint from my hands. Last night's slight frost, half-thawed, glistened among milk-colored patches of still frozen canvas boat covers all over the Yard. One month ago, as seen from above, Billings', or "the Yard," would have looked like the site of a Boy Scout Jamboree. But the acres of tented canvas stretched tight to shield occupants from weather now had gaps in it, campsites vacated by the more serious and gung-ho lobster fishermen.

The old-timer next door was one of many who had his boat ready before I did. He stood proudly in the middle of her deck, arms folded across his chest, while the two giant straps of the travel lift were placed at about one-quarter and three-quarters of the boat's length. "You're a mess!" The old man had yelled down to me over the din of the engine driving the boat-lifting and -moving machine.

"I'm not a very neat painter. I hate rolling the bottom. I've paid my penance to the lobster god!" I declared with a smile, stretching my arms out to embrace what little warmth the sun had to offer.

"Paying penance? Are you in arrears from last season?" he asked.

"Hell, no! I'm paying in advance for the bounty I am sure to receive this season."

". . . paying for something you may not get."

"I'll get it . . . I hope." Realizing that I sounded less than optimistic, I added, "Ask and you shall receive! That's in the Bible, right? Well, consider this asking."

"Yes, the Holy Bible. John, Acts. I don't think it was said in reference to lobsters though." And away he went, held in the palm of his boat, cradled by the straps to the edge of the water where he was slowly lowered and set free from the labor of the Yard.

The launching of the *Mattie Belle* had been scheduled for "after lunch," and I had to hustle at that point to unwinterize her before the splash. I had set off in search of a ladder with which to climb aboard. Boatyard ladders seem to have lives of their own. There was some form of ladder propped up against nearly every boat in the Yard except mine. Boat owners do not possess ladders, boatyards do. And Billings' has one of the finer and more eclectic accumulations I have seen. It is always tempting to snatch the ladder from the boat nearest yours. But as no two boatyard ladders look alike, stealing one would never go undetected and would no doubt leave the original users more than merely stranded.

It has been my observation that in a boatyard almost nothing is thrown away as useless. Old scraps of wood, broken pallets, and crates are left lying around to be put to use in an assortment of projects. Blocks and boards of various widths and lengths are needed everywhere; the welding shop, the paint shop, the woodworking area, even the area where the diesel mechanics work, all have a pile of well-used, but useful, lumber. The rotten, cracked, gnarled, and overly weathered fragments that have simply degenerated beyond imaginable use are bound together in a ragtag fashion and labeled "ladder."

In the faraway southwest corner of the Yard, a rickety specimen lay beside a boat-shaped wet spot in the gravel, obviously abandoned just this morning. The ladder was taller than I required, but since it was the only one available I started dragging it in the direction of the *Mattie Belle*. The rungs consisted of broken broomsticks, driftwood, and pieces of two-by-four. "Now that's a beautiful thing," yelled a coffee-breaking worker from a picnic table as I raked the ground in front of the paint shop.

Ready for a breather, I had stopped to talk with the familiar faces I hadn't seen since last November when I put the boat up for the winter. "Me? Or the ladder?" I asked. I ran a hand through my hair, a combination of knots and curls, some of which were stiff with bottom paint, surely disturbing any hairstyle I had acquired after lopping off my ponytail weeks before. My hair had become a barometer, indicating relative humidity or dew point or fog or anything moisture-related. Mist caused tendrils curled so tight that I had been questioned about permanents. Hot and humid was forecast by kinks and frizz. Meandering waves and rambunctious twists were my favorite pattern, brought on by the conditions of late fall. Today's weather had created somewhat of a tangled mess, of which I was suddenly quite aware.

I liked the employees of Billings' Diesel. There was a core group of full-time year-rounders who were quietly competent, a trait that many Mainers share. Any person's particular talent can be greater appreciated when it is revealed unexpectedly rather than boasted and half proven. The group of middle-aged men who held down the picnic table possessed the refreshing ability to surprise and impress by doing a variety of jobs well while appearing to enjoy their work. Among this crew were a jumble of riggers, painters, fiberglassers, woodworkers, mechanics, and heavy equipment operators. Onlookers would never have suspected these gents of being among the best in the business, but they were.

We shared banter about various winter activities, and the conversation quickly came to questions, predictions, and prophecies for this year's lobster season. All residents of coastal Maine, whether they live on an island or the mainland, are part of the fishing community and have a degree of interest in the progress of each season. Billings' customers consist of owners of both work- and pleasure boats. Many of the workboats are employed catching lobster. Harland Billings, the owner, also runs a lobster-buying operation and knows the importance of keeping the fishing fleet up and running. Someone from the table remarked, "My brother-in-law hauled two hundred traps for seventeen lobsters yesterday." Everyone present was in agreement that early spring was looking worse than usual for the fishermen, quoting numbers amazingly large and depressingly small of traps hauled and pounds produced.

"I guess you're not in any hurry. You're certainly not missing anything. Not yet anyway," someone suggested, noting that I had not moved one inch closer to the *Mattie Belle*. Cigarettes were crushed out, and soda cans and Styrofoam cups were dropped into a trash barrel as if some whistle had sounded, beckoning the men back to work. Reminding myself of the hours and days of work ahead of me before I could report any bulk of lobster landed, I dragged the ladder the remaining distance. My pace was brisker now.

I placed the ladder against my boat with the stoutest of its rungs near the ground and gingerly ascended. Standing safely on her deck, I secured the ladder to a ring bolt on the boat's gunwale with a piece of pot warp (line or rope used in lobster fishing), more for safety than fear of ladder thieves.

A couple of hours and two bloody knuckles later, I was nearly ready for the launch. The pair of 12-volt batteries under the deck had been hooked up. Terminals and posts had been scraped clean, and all cells had been topped off with distilled water. From an access hole in the forward bulkhead, I had attached the end of the 2-inch hose to the thru-hull fitting supplying raw water to the engine's heat exchanger, or

29

cooling system. All antennas had been put back into their upright position after lying horizontally all winter. The antenna job had resulted in the only long and loud string of profanity to escape my mouth, as I could not find a $9/16$-inch wrench, and resorted to slightly rusted vise grips. I had checked the engine's coolant level and was inspecting the dipstick for lube oil level when I heard the travel lift approach.

The lift operator, Bar, caught my eye with a questioning look. Assuring him that I would be ready to be launched by the time he carried the *Mattie Belle* to the water's edge, I hurriedly checked all hose clamps. The bearded man in the blue plaid shirt rigged the giant straps around the belly of the *Mattie Belle* and climbed back into the lift's driving and operating station.

It was a slow and smooth ride with rubber tires absorbing gravel peaks and muddy valleys. The boat swayed shallowly, almost indiscernibly, back and forth in the swinglike straps of the travel lift. I always enjoy this short ride to the Yard's launch site. It brings with it a mix of feelings of accomplishment for completing step one in the lobster season's preparations, and anticipation of steps two and beyond. If the engine starts with one turn of the key, I thought, I will be greatly relieved to bid farewell to the Yard until next winter, when I would return with stories of the most bountiful season ever. It's a romantic moment until real thoughts of hard and monotonous work creep in and overwhelm.

Floating, with no signs of leaks, was the first "phew." Clicking the battery switch "on" with no showering of sparks was the second, and the engine starting just as it had the morning I left the Island so many cold days ago provided the final relief from worry — for the moment.

So here I was on the *Mattie Belle*, with my two garbage bags full of stuff, thinking back to the hours of preparation I had just done and looking forward to the months of lobstering ahead. I waved good-bye to Bar, mouthed a silent thank-you, pulled the gearshift into reverse, and backed the *Mattie Belle* out of the straps that now lay limp beneath the boat's hull. I then turned on the radar that I would certainly need to pick my way through the fog for the 7-mile trip to the Island.

I heard someone yell "Hey, Linda" just as I swung the boat around toward home. It was from a bit of a distance, but even with the poor visibility I recognized three of my favorite Billings' employees on the deck of a beautiful sailboat waiting to be slipped into the water. I waved and smiled, happy to be heading to the Island in my pretty little workboat. As if on cue, the men turned their backs to me and dropped their pants, exposing three snow-white bottoms. Three middle-aged men had mooned me a fitting send-off, I supposed. I laughed all the way to Merchant's Island, where the lobster buoys were already so thick I had to weave in and out of them to dodge their warps with my propeller.

Now was the time for the real work to begin. The presence of all the multicolored buoys that appeared one at a time through the fog was like a cold slap, alerting me to the fact that I was already behind in my preparations, in arrears to the lobster god. I noted the arrival of three new buoy colors to the area, and wondered what would be done about them. Would this be the season my fellow Islanders would follow through with plans to alleviate some of the pressure applied by those who encroached on our territory? There had sure been a lot of talk over the last four years, and each season the pressure to do something about the interlopers increased. Eventually something would have to give.

By the time I reached the end of Flake Island the fog had settled down low. My first view of home was the hilltops poking through the gray vapor; the bulk of the Island was still hidden. The fog was proving moderate and passive, and it insulated all the senses — a film over my eyes and cotton in my ears. In the four years I've been back I had learned that fog, like all weather, could come and go quickly and without warning, even if I could not.

THE LOBSTER
ASSOCIATION

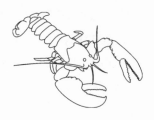

If it weren't for the mosquitoes and blackflies, I could honestly say that I enjoy working in the trap lot, the place I store my lobster traps and fishing gear. There is great satisfaction in methodically moving virtual mountains of gear from one pile to another, from the "undone" to the "done," and watching the mountains shrink and disappear as the gear is taken away by truck, destined for the bay. Anticipation builds as the trap lot empties, for every trap moved from land to sea will surely fill again and again with soft-shelled lobsters. Working the trap lot is similar to splitting and stacking cords of firewood. It's labor-intensive, and you can see results at the end of the day, but you're not making any money. As last year's proceeds dwindle, anticipation of the new season begins. "I have to get my traps in the water" is the anxious cry of all gear-prepping, buoy-painting lobster catchers — with emphasis on "have."

An unknowing passerby might think my trap lot a dump or junkyard, mainly due to its appearance. While there is no garbage odor or vicious dog, it is impossible to keep heaps of fishing gear tidy and organized. The gear I cherish most, fishing rods and accessories, lives within the protective walls and roof of a workshed, while those things not susceptible to salt air and weather remain scattered around the property.

Glancing around the area for the first time this spring, I was both pleased and discouraged with the inventory. Bundles of buoys and toggles hung like bunches of grapes from branches of trees. Blue tarps covered piles of coiled 20- and 30-fathom pieces of multicolored line. A small sheet of canvas, wrapped cocoonlike around a deflated rubber raft, held brown spruce needles in its nooks and crannies. The raft hadn't seen the light of day in so long, I figured it must be mummified by now. Two old plastic bait barrels sat relaxed against trees. The barrels were full of assorted garbage that half-floated in stagnant water. I wished now I hadn't used the barrels for garbage. I wished I had not allowed them to fill with water and sit for several months, fermenting. And I wished I had not been so impressed last fall with the fact that John Cousins, who was helping, could stack lobster traps six high. John was not around now, and I could only reach the fourth one. I wished I had coiled my pot warps neatly rather than bailing them haphazardly into the "trailers." I wished I had cleaned my buoys more thoroughly (or at all) before retiring them for the winter. Most of all, I wished that I had started the gear work prior to this season's first bloom of flying insects.

Trying to ignore the mosquitoes, I got to work and was soon joined by my father. My father and I work well together. Amazing, I thought, that my dad, who had spent so many years urging me to "get a real job," and discouraging me from taking my college diploma to sea, had now retired from his "real job" only to find himself fishing with me. I was sure neither of us had planned for it to happen, it just sort of worked out that way. Dad was quick to catch on to the fact that the possession of fishing gear shifts back and forth from "my" to "our" depending on its condition — for example: My gear is in good shape; *our* gear needs work. We were in agreement that we wanted to begin setting traps as soon as possible, and in order to do that, six hundred traps, three hundred buoys, three hundred toggles (small floats), and miles of coiled line needed to migrate from the "undone" pile to the "done" pile.

I had known from the start that I would need a crew of one to work in the stern of my 35-foot lobster boat — a "sternman." I knew I didn't need any witnesses to the many blunders I would no doubt make while learning how to catch lobsters, but I would need help. So the ideal person would be someone who knew less than I did, who would not tell me what to do aboard my own boat, and would do as he was told without question, no matter how ridiculous. Someone with discretion. The natural candidate was a retiree with ties to the Island going back four generations. My father.

Today, Dad joined me in the trap lot and began prepping buoys for paint while I overhauled lobster traps. The obstacle of trapstack height was easily overcome. I grabbed the fourth of each six-trap vertical stack, pulled, yelled "Timber!," and jumped out of the way as the traps crashed to the ground where I could easily reach them. The first "timber" and "crash" solicited one raised eyebrow from my quiet father. After that, he didn't look up again from his work scraping dried plant life from the buoys despite the continuing clamor.

Lobster traps are rectangular boxes made from vinyl-coated wire mesh material called "trap wire." I fish both "3-footers" and "40-inchers." The names refer to the length of the trap. Many lobstermen prefer "4-footers," but the extra 8 inches add too much bulk and weight for my back to handle. (Four-foot traps are also known as "chiropractic specials.") I'm not sure how much my traps actually weigh, only that they gain weight as the day grows old. My traps are all 16 inches high and 23 inches wide. They're all "three-brickers," meaning that each trap has, within it, three masonry bricks, the purpose of which is to distribute weight in the trap to ensure that it lands right-side-up when set, and not wander the ocean floor in strong tides and lousy weather.

The "heads," or openings, in my traps, which allow lobsters to enter and move from "kitchen" to "parlor" (two compartments of traps), are made of twine, netting, or webbing and are cone- or funnel-shaped. The trap heads are "shrimp mesh," plastic twine sewn into netting of 1-inch mesh. Older traps have heads made of nylon line, and a much bigger mesh size. They do not fish as well as the shrimp mesh heads. I have no idea why.

All trap elements are inspected in the overhauling process each spring. Repairs are made before the trap is set, or launched, since it is easier to work on land, even with the voracious mosquitoes, than it would be aboard a boat. I was happy to find the bricks secure, heads in one piece, and doors, which are on top and allow access to the interior, functional in the traps I handled this morning. I replaced the biodegradable hog rings (fasteners) in the escape vents on the doors of the traps, snapped this year's tags into place, and overhauled all lines. Slowly there emerged a "done" pile.

All traps are equipped with hard plastic escape vents that have oval openings large enough to allow "short" or undersized lobsters to exit a trap at will. Each of my traps has two vents, one in the door and one in the parlor end. Maine State Law requires that one vent be secured with biodegradable hog rings, while the other may be set with stainless steel, requiring little or no maintenance. The idea behind the mandatory biodegradable vent is to ensure the liberty of all lobsters within a trap that may be lost or neglected. "Ghost gear," or lost traps, are not a threat to lobsters' lives because the biodegradable hog rings deteriorate within a season, allowing the plastic vent to flop open, leaving a large exit. All biodegradable rings or remains of rings must be replaced when overhauling traps if a fisherman expects to catch anything. Otherwise, lobsters will find open vents, and fishermen will haul up empty traps. I was clumsy with the hog-ring pliers at first, but found more ease and comfort as the morning progressed. "Are you going to the meeting tonight?" I asked my father, referring to the monthly meeting of the Island Lobsterfishermen's Association.

"One of us should go," he said.

"It won't be me. I'm not a member."

"You would be if you attended meetings."

"Forget it," I replied with a smile, knowing how much my father hated the gatherings. In order to be eligible for the "annual bonus," fishermen must be members of the Association. In order to be a member in good standing, fishermen must be voted in by a majority of members and attend at least six monthly meetings each year. "You'll split *our* bonus with me, won't you?" I asked.

"You mean *my* bonus. You're not a member."

That was the end of the conversation.

There was a lot that had not been said, a lot that was not necessary to rehash. I enjoyed replaying silently in my mind my initial involvement in the Association. I remembered clearly my inner turmoil in deciding to accept the invitation to attend my first meeting. I had not, four years ago, come to the Island to join anything other than the community. I had no intention of sharing ideas or resources. I had no desire to become a member of any organization, association, club, or committee. But I did want to fish among, and be accepted by, the Island lobstermen, all of whom attended meetings of the Association.

I remembered debating whether or not to attend right up to the door of Billy Barter's shop at meeting time. I recognized most of the vehicles parked on the road, and all of the voices coming from within the shop. I opened the door and was welcomed by all. Buoys that hung from the ceiling of the workshop looked like red-and-white helium-filled balloons and filled the air with the smell of fresh paint. Folding metal chairs and an assortment of straight-backed wooden ones were arranged in a circle around the perimeter of the wide-planked floor. Most of the seats were occupied by fishermen waiting patiently with hands folded in laps for the meeting to begin.

I couldn't help but wonder, actually worry is closer to the truth, what kind of reception I might get from the men who traditionally fished around the Island. I knew all the stories about sabotaged boats and gear, and the tales of personal injury experienced by unwelcome parties up and down the coast. The last of my hard-earned savings was wrapped up in my boat, traps, line, and buoys. If I lost them, I would have no choice but to admit failure and go back offshore. How would the Island fishermen feel about sharing? Lobster fishermen all along the coast of Maine had been thriving in a booming economy, the key to which was a renewable resource, a lobster stock that had proven to be resilient beyond what a reasonable person could fathom. This had been one basis for my decision to join the multitude of newcomers to the industry. And I hoped the locals would consider my family name as entitlement to a sliver of the pie.

I understood from what I had gathered prior to the meeting that the Association was a co-op of sorts, an outlet for the sale of lobsters and a purveyor of bait for the Island fishermen who were members. My years of commercial fishing had taught me that any "fishermen's organization" is an oxymoron in that groups of fishermen are, by nature, unorganized. Through the years I had attended hundreds of meetings of offshore fishermen, ranging from formal gatherings to impromptu coffee-shop discussions, and all had a common denominator: disagreement. Fishermen are necessarily independent in thought and action, so it's common for a discussion to turn to disagreement, and a disagreement to escalate to a shouting match, which occasionally results in fisticuffs in the parking lot. It is a virtual impossibility to find two fishermen who can come to total agreement on any single point other than the price of whatever species they are harvesting being too low or diesel fuel too high.

So there had been no question in my mind that this meeting of the Isle au Haut Lobstermen's Association would be total chaos, out of which nothing would be accomplished other than a few insults and possibly some hurt feelings.

My first impression, as I looked around the room and waited for the shouting to begin, was . . . day care. Why would anyone bring children to a meeting of rowdy, crude-talking fishermen? There were half a dozen small children in attendance: two squirmed in their fathers' laps; two sat in their own chairs looking quite bored; and two others were sprawled out in the center of the ring of chairs with coloring books, where I was sure they would be trampled by the first men to scramble out the door to settle their differences. My second first impression, if there is such a thing, was . . . sewing circle. Two elderly women sat closest to the wood-burning stove with small canvas bags at their feet, out of which stuck the ends of knitting needles and balls of yarn. So far this did not even remotely resemble any assembly of fishermen I had ever attended or heard of.

Both curiosity and confusion climaxed when Al Gordon entered the shop, apologized for being late, sat down with an open notebook, and proceeded to read the minutes from the previous meeting. Minutes! While I sat with my mouth open, a motion was made, seconded, voted on, and passed to accept the minutes as read. I marveled at the strict parliamentary procedure. I was beyond surprised that Al Gordon, one of the Island's general contractors, was secretary of the Fishermen's Association. Al wrote furiously with a pen he gripped with the stubs of fingers remaining after several mishaps with a table saw. Belvia MacDonald read the treasurer's report. I may have been in shock. I had heard the term *gentlemen fishermen,* but until then had thought it a joke of some kind.

Two of the kids drifted off to sleep, and Billy Barter snored sitting straight up in his chair with his arms folded across his chest. All business was concluded, and the meeting moved to a less formal format of polite discussion. The discussion heated up to lukewarm, by my standards, when Dave Hiltz brought up the topic of unwanted newcomers fishing the Island waters from the mainland.

A dark-haired and good-looking man in his early thirties, Dave and his wife, Debra, were relatively new to the Island, having moved here from Bar Harbor six years ago. Their move was made possible through the ICDC program, Island Community Development Corporation. The ICDC is intended to increase the Island's year-round population by providing acceptable newcomers with affordable housing and land on which to build their own homes. Proponents of the ICDC program wish to strengthen the fishing fleet and keep enough children on the Island to maintain the one-room schoolhouse education of K through eighth grade. Dave fishes hard and has a daughter, Abigail, making him and his wife good choices for repopulating. Dave and Debra are hardworking people who have adjusted well to Island life, while other couples had come through the ICDC program but found the winters much too long. Dave immediately became my dearest fishing buddy on the Island. We share a love of boats and fishing, seldom speaking of anything else. Dave is a guy who will drop whatever he's doing to do someone a favor. He had a lot invested in his move to the Island and was quite vocal about what he thought needed to be done in regards to protecting our fishing grounds. At this meeting of the Association, Dave got quite excited as he reported sightings of a number of "big, new boats" setting traps in the area. As Dave spoke, I watched Jack MacDonald.

This was the first time I had seen Jack with a beard, and thought he looked like Abe Lincoln. Growing up, I had known Jack as the lobster fisherman father of my friends Danny and Nita, and had as much contact with and knowledge of him as most teenagers do with their friends' parents — not much. Knowing what little I did about coastal fishing, I understood that Jack Mac-Donald would be the most helpful and important figure in my transition from off- to in-shore. Jack was the head of the Association and a leader in the community, working for years in town government as selectman. He listened intently as others chimed in with mutual sentiments of disgust for the steady increase of pressure on their fishing grounds from outsiders.

Everyone was in agreement that something needed to be done, and there were clearly two different opinions on what would remedy the problem of overfishing. Solution number one: Engage in a "gear war." Simply cut the buoys off any unwanted traps that appear in the area we Islanders would like to protect for ourselves. Of course, any fisherman worth his salt who suspects he has lost traps in this fashion will certainly retaliate by cutting someone else "out of the water." Cutting would more than likely escalate to other types of destruction, and everyone would lose except for the trap manufacturers. As one fisherman had pointed out, sabotaging another fisherman's gear is illegal, punishable by loss of your fishing license and possibly jail time. He made it clear that he would not be involved in any gear war.

Solution number two: Push a bill through the state legislature to form a conservation zone around the area that Islanders traditionally fish. This would amount to an exclusive fishing zone governed by more restrictive regulations than the rest of the state waters, a zone to conserve the lobster resource and preserve the Island's year-round community. Similar plans had been in effect around the islands of Monhegan and Swan's and had, by all accounts, been very successful. The downside of the legal solution would be time. The results of a gear war would be felt immediately. Going through the steps and red tape involved in passing legislation into law would be tedious.

Quickly and quietly a unanimous decision was reached to pursue solution number two. I was convinced that the men were genuinely concerned for the lobster resource and the Island community. I was also aware that outsiders would see this action as a move to stake a claim on a productive piece of bottom, greedily keeping the competition out, decimating the lobster stock all by ourselves while lining our pockets. All members of the Association seemed committed to the idea in spite of Jack's constant reminders that it would be a long, hard, uphill battle. Finally it was decided that Jack should do some research on how to go about getting our own fishing zone, and that we should meet again before an entire month passed. There was a sense of urgency.

As my first and second lobster seasons progressed, I attended monthly meetings and a number of special meetings, all of which had one thing in common: Each meeting required a vote of all members by show of hands to approve and support the work of Jack and a few others in moving toward legislation to create our own fishing zone. In two years of meetings, discussions, research, and letter writing, there was never even one single hand raised in opposition to the effort being made in the name of preserving our precious piece of America and protecting our livelihoods. I had wondered why Jack felt it necessary to constantly hold these votes. Was anyone really apt to have a change of heart or develop cold feet? Who would give up a fight for a way of life? I would later find out the answer to that question.

So, over time, I simply stopped going to the meetings and let my father serve as my proxy, leaving me with no bothers other than the mosquitoes. My father would continue to keep me posted about anything I might want or need to know with regards to the Association, including the meeting he would complain about attending tonight.

I kept track of the number of traps completed by the number of trap tags I used. I wore a three-pocket carpenter's apron across my abdomen. One pocket held hog rings and pliers, one held a knife, and the largest held the plastic tags mandated by the state of Maine to be snapped into each trap. Each tag looks a bit like a miniature hospital identification bracelet. Tags are applied for when lobster fishing licenses are renewed, which happens annually. Tags are stamped with a fisherman's license number (mine is 936) and the zone in which a fisherman chooses to work (my tags say "zone C"). Tags are also marked with the proper year and are color-coded from one year to the next. Tags are priced at twenty cents apiece, with a maximum limit of eight hundred tags per licensed fisherman. The tag limit is part of the "reduction in effort" in all fisheries. Both the state of Maine and the fishermen hope that by limiting the number of traps in the water, lobster will not be overfished.

"Hey, Dad, do you think these old tags are number two plastic or household trash?" I asked, contemplating a new recycling nightmare.

"Leave them in the traps, and we won't worry about it. I have fifty buoys cleaned. I'll take them into the shop to start painting. Mosquitoes are fierce . . ."

"No shit. They're driving me nuts," I said, swatting the air above my head. My father dragged the buoys he had scraped and brushed and rubbed with course sandpaper through the shop door, which he was quick to close behind him. Fifty buoys would be enough to set one hundred traps. I was falling behind. According to the number of tags I had used, I had completed only forty traps and had not begun to go over the lines that I had been removing from the trailers and throwing into a heap. Desiring an inside job, too, I dragged the pile of sort-of-coiled warps into the shop, another mountain to move from undone to done.

What I needed to do now was make sure the warps, the lines that connect the traps to each other and to a buoy, were the right length and neatly coiled. There is a lot of measuring and splicing involved because over the course of a season the lines tend to get chewed up (often by propellers), and it's usually easier just to tie them back together and worry about them the next year.

Basically, one line connects the main trap to a trailer trap (this is usually about 60 feet long) and another connects the main trap to a buoy. Obviously, the two traps are supposed to lie on the bottom of the ocean and the buoy should float. It's important to get

51

the length right for whatever depth of water you're going to fish, otherwise the traps will suck the buoy below the surface, resulting in lost gear.

I measured, cut, spliced, and replaced carefully as my father brushed bright orange paint over last year's coat, buoy after buoy. Dad hung wet buoys from the shop's rafters to dry overnight. Tomorrow, stripes would be added. The protective coat of Filter Ray, which protects the paint from fading in the sun, would then be applied, and would also need a day to dry. By the time these first fifty buoys were ready, I would have a hundred traps with lines ready, too.

The buoy color printed on my Maine State Lobster License is "orange, yellow, white." The orange I use is "blaze orange," and the type of paint is Day-Glo. It's quite bright and shows up better than some others in fog. The white and yellow stripes, one of each around the butt end of every bullet-shaped float, are bands of colored vinyl or electrical tape. The loose end of each strip of tape is kept from peeling away from the buoy with a 1-inch galvanized nail with a large head. When my mother and Aunt Gracie painted buoys for me, the stripes were painted. But they quickly learned that buoy painting is monotonous.

My Tom Sawyer routine did not work the second time I tried.

I continued working with the line, and my father continued stroking buoys with blaze orange. Mindlessly cutting and splicing, my attention was suddenly taken by the screaming coming through the trees from Frank and Rita's place. The noise of the couple arguing was nothing new, and Dad managed to ignore it. It sounded like Frank had locked Rita out of the house this time.

At the end of the day, fifty buoys hung from the rafters drying, and forty traps, with warps perfectly coiled and neatly tied, waited to be set. We were anxious to get some gear in the water, but there was so much more prep work to do. All conversation in the trap lot was weighted with anticipation of catching lobsters and hoping my checking account would not dry up before realizing the fruits of our labor. Slowly the piles shifted from undone to done. And eventually Rita convinced Frank to let her back inside.

RITA

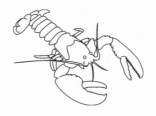

Some time ago on a lark, I had my tarot cards read, and immediately wished I hadn't. I had always been skeptical of anything even remotely mystical, and had sat down across from the reader, who had a milky film clouding one eye, with an understanding that his act, although pure hoodwink, could be fun and nothing more. And at first it was. The oversized cards painted quite a rosy picture of my future in both love and money. Everything was perfect, and I was considering a healthy tip in return for the man's genius, then his focus shifted from the cards to my face. In an almost scolding tone he told me that I "often put people up on a pedestal who do not belong there," and that "others regard perpetual goodwill as foolishness." I didn't like being called a fool and wondered what combination of cards indicated this. I left the table believing I had some major shortcomings. At the time, I couldn't think of a single person I genuinely disliked, and thought the reader or the cards quite perceptive. It bothered me, for some incomprehensible reason, that I didn't really dislike any individual or group of people, and I vowed to work on this.

When I first met Rita, I was unaware that she had the potential to be the first person I really disliked, although she clearly had an ability to bring out the worst in other members of our Island community. My introduction to Rita (she prefers to be called Wilma but most people do not oblige her in this) was while walking toward Moore's Harbor with my mother. It was one of those idyllic summer days, of which there are so few — just the right temperature, not too hot, not too cool, and air so still and clear it seemed to not be there. Mom and I, walking for exercise, shared a comfortable silence and a mutual contentment. Rounding a sharp corner, facing a slight upgrade, I raised my eyes from the road, which demands carefully placed steps to avoid turning an ankle, to see an eccentrically clad elderly woman pushing a bicycle toward us. It seemed, at the closing distance, she was wearing far too many clothes for such a temperate day. "Oh, great, it's Rita," my mother said in a tone indicating anything but delight.

"Who's Rita?" I asked as we drew near.

"I'll tell you later," Mom whispered. "Hi, Rita."

Although the bike was lacking the telltale streamers from the handgrips, and rubber ball horn, it did sport the obligatory basket overflowing with "stuff" concealed by what was left of a tattered, green plastic garbage bag. If I hadn't known such things do not exist on the Island, I would have guessed that Rita was on her way to or from a homeless shelter. All that was missing was the shopping cart. But she did have the bike, which she pushed rather than rode, another dead giveaway that all was not shipshape topside. The woman stopped, wanting to talk. Smiling, she revealed an oblong of pink gums encircling a black hole framed by dark, beady eyes and the type of chin made prominent by the absence of teeth. "Oh, hi, Martha. Who is that with you?" she asked, as if I were not standing directly beside her.

"This is my daughter, Linda. Linda, say 'Hi' to Rita."

"Hi, Rita."

Rather than a polite "Hello, nice to meet you," I received a long, hard stare from the woman I had already determined must be crazy. She was conducting such a thorough inspection of me that I began to feel uneasy, wondering what she was thinking. When she finally spoke, she put an index finger very close to my chin and said, "You have a pimple right there." Rude, I thought. I had forgotten about the zit, and now that Rita had reminded me of it, I felt self-conscious. Now my mother was also looking at my chin and awaiting my reply. Was I to thank her? While I struggled for an appropriate response, Rita wheeled her bicycle around, so that the three of us traveled in the same direction, and said, "I have something for pimples at home." Rita pushed with a stride that had suddenly taken on some purpose. Two steps ahead of my mother and me, she complained in a stream-of-consciousness monologue about a particular brand of ink pen that was making her seasick every time she used it to write a letter during trips back and forth to the Island aboard the mailboat. I would have suggested that she not attempt to write while on the water, but she was careful to not leave a space into which I might interject. She never paused. I rolled my eyes at my mother, who smiled and shrugged.

We stopped in the middle of the road in front of the house Rita shared with her ex-husband, Frank. Frank is a kindhearted man who must have done something terribly wrong to deserve Rita, my mother quietly explained while Rita entered the house to retrieve some of her zit remedy for me. Apparently Frank went away "on sabbatical," and returned with Rita in tow. The two married, but soon divorced when it was discovered that the social benefits from the government were better when received by two individuals rather than one married couple. As for the pair, the coupling, by all reports, had never been blissful, as evident by the not uncommon locking-out of the house of one party by the other. This banishment was always accompanied by an enthusiastic thrashing of the front door by the outside party, and shrieking and screaming from both sides of said door. My mother also explained that Rita would expect to be paid for whatever she produced, and that perhaps a dollar would be appropriate.

"You're kidding. I should pay that nutcase for something I didn't ask for and don't want?"

"Just give her a buck, and we'll get out of here."

"Jesus . . ." I dug in my pocket, and pulled out a crumpled-up dollar bill just as Rita dragged her rubber-booted feet back to where we dawdled in the road. She handed me a small plastic prescription bottle that contained some type of goop or cream, grabbed the dollar, and shuffled back toward the house, where Frank had appeared in the doorway lighting a pipe.

"Hi, Martha. Hi, Linda," Frank waved and called pleasantly as my mother pulled me down the road by my arm.

As soon as we were out of earshot, I asked, "What's with that guy's hair?"

"Frank shaves his own head. Doesn't do a very good job."

"I guess not. Kind of patch-worky, like he got interrupted and didn't quite finish."

"Rita must drive that poor man nuts," my mother sighed, with a clear affection for Frank.

"Well, yeah . . ." I laughed, "someone is driving somebody nuts there."

"Aren't you going to try that cream?" Mom asked with a grin.

"No! I'm not putting any of this shit on my face."

"Why not? You do have quite a pimple, and Rita's complexion is flawless."

My next two encounters with Rita didn't do much to reassure me of her sanity, nor of the manners of some of my family members.

While I was sitting with my cousin Diane on her front porch, Rita approached from the main road. Three steps into the driveway, Rita was greeted by Diane bellowing something to the effect of "Rita! Get the hell out of here!" Rita ignored this and trekked closer to the porch, where Diane now sat on the very edge of a plastic chair, appearing to be ready to pounce. "What do you want?" My cousin's voice barked with a newfound coarseness.

Rita peered from the small opening of the scarf with which she had shrouded herself from the shoulders up. Her pleasant reply, somewhat muffled by the layers of cloth, I understood to be "I really love your earrings, Diane." I was sure I had not heard correctly until Diane batted the compliment to the ground with a knee-jerk reaction of assuring Rita that she could not have the earrings or anything else, and that Rita should leave the premises immediately if she knew what was good for her. Rita asked politely if she might use the telephone before going. Diane's reply was studded with colloquialisms that I was accustomed to using myself only while aboard a fishing boat. Every four-letter word coming from my cousin's mouth seemed to poke Rita in the eyes; she received each one with a frantic blink and flinch. I fully expected Rita to say "ouch" when Diane drew a breath. But she silently shuffled back toward whence she had come, stopping only briefly to inquire about the possibility of borrowing a spool of thread. The answer was of course, "No fucking way."

Diane watched Rita disappear behind some trees at the corner of her driveway, shook her head, and said pleasantly, "Oh good. She's going to Lucinda's." The idea that Rita would now bother Diane's closest neighbor, with whom my cousin enjoys a long-running feud, appeared to amuse her, and she chuckled. I thought Diane had been extremely cruel to Rita, for whom I was almost feeling sorry. My cousin clearly sensed that I had been put off by her harsh treatment of Rita and went into a brief explanation, which I read as a warning: "Rita is crazy like a fox. She is the queen of manipulators. She is clever and insightful and will prey upon your weakness, your desire to be nice to everybody." There it was again. I looked around for the tarot cards.

By early fall, I wished I had heeded the words of advice from Diane and my parents. (My mother had suggested, on more than one occasion, that I get up in the morning, look in the mirror, and practice saying "No,

Rita.") Rita was at our door every day asking to use the phone or to borrow some item that she always received and never, with the exception of a pair of underwear, returned. I had to consider the possibility that Rita had succeeded in totally wrapping me around her crooked fingers or driving me crazy, when I arrived home late one night from a party and saw through a window that Rita was there, uninvited, watching television. (My parents were off-Island at the time.) I wanted so badly not to deal with her at this late hour, not to risk ruining what had been a lovely evening, that I crept through the woods to my cousin's house. I sneaked up the staircase and into a guest bed, leaving Rita undisturbed and comfortable in our house, where she no doubt made several phone calls. I had actually reached the point of preferring to pay Rita's phone bill to listening to her nonsense myself.

The final straw is one that I hesitate to share, but will in order to fully paint the picture of what Rita's presence does to otherwise sane people. Entering my parents' house one day through the sliding glass doors separating the front deck from the living room, I heard my father yell, "Get down!" Both mother and father hit the floor with the speed of people living in a war zone. My father waved his hand frantically at me, motioning me to lie on my belly on the hardwood floor, which I did. Dad crawled like a veteran of the front lines over to a window overlooking the boardwalk that stretched between our driveway and back door, peeked out, then ducked back down quickly, covering his head with his fore-arms. In a whisper now, I heard him ask my mother, "Did you lock the door?"

"I didn't have time" came the answer from my mother, who had managed to squeeze behind the couch. I slithered over to a window, curious as to the danger, and looked out to see . . . Rita. She walked slowly, stooping to pick up a Blue Willow dinner plate that had been left by Mom full of cucumber butts for the deer. Rita slipped the plate into her oversized shoulder bag and headed toward the next thing that caught her eye.

"She's going for your chain saw, Dad."

"She'll only take what will fit in her bag."

"Well, it's a pretty big bag!"

A partially stifled snicker came from be-hind the couch, and triggered a gale of laughter that stopped just as abruptly when the three of us feared being discovered by Rita, who continued to prowl around the yard. Rita eventually disappeared, with only the plate and a small garden trowel. As my mother crawled out of hiding, Dad and I enjoyed the freedom of standing in full view of the windows. I brushed some dust from my jeans and said, "I can't believe that hiding in your own house is preferable to actually dealing with Rita."

"The only way to deal effectively with her is to be cruel. We prefer avoidance." I realized that I had come, quite naturally, by my propensity to be me. It was in the DNA.

I remain mystified by the tarot card reader's ability. As it turns out (and this only adds fuel to my discomfort in her company), Rita has some degree of talent as a prophet herself. I am aware of several instances when her babblings were shrugged off as dementia and later fell into the category of "gospel truth." The best example of Rita as seer began one fine day on the town dock before a large audience. Rita was pontificating with some volume and pointing an accusatory finger at one of our upstanding citizens of the female gender. Rita seemed bent on exposing this woman as an "evil temptress" and "seductress," and even went so far as to announce to everyone within earshot that "she may look innocent, but she is sleeping with other women's husbands." The target was naturally very embarrassed, but she seemed able to laugh it off with the rest of us as one of Rita's crazier moments.

A few months after this particular Rita outburst, our upstanding citizen of the female gender (who was very married) was caught "in the saddle" by her close friend — and wife of said "saddle" — at a late-night rodeo on top of the table at our tiny public library. Some people were impressed that Rita had predicted this. One of the things that I do love about my fellow Islanders is that it did not seem to occur to anyone that Rita might not be a talented soothsayer but was, possibly, a more common, garden variety Peeping Tom.

SETTING OUT

It would take us three days to move the first hundred traps from the lot to the water, setting just thirty traps each day on the high tide. I wished I had a dock on which to store and overhaul my gear, and from which to set out, eliminating two handlings of the traps each season. Rather than loading the traps into the back of my truck and unloading them at the town landing, I could then simply move the stacks directly from dock to boat. But I didn't. Hence lots of loading and unloading.

Dad and I once more loaded the back of my 1983 Dodge Ram pickup with twenty-four traps, coils of line, and buoys, and headed to the town landing. Looking out over the hood from behind the steering wheel, I thought someone should have applied Filter Ray to the truck's paint years ago; it had faded from what I imagined was once a dark maroon to an unattractive rose. The only fresh paint this truck had seen since birth had been in the form of balls fired from guns in last summer's paint-ball war, when the truck had become my team's "battle wagon" and had been pelted with balls of every conceivable color from semi-automatic weapons being fired from behind trees, rooftops, or other moving vehicles. The war was great fun until people started getting hurt. Oh well, that was last year's activity. We would be doing something new this summer, I thought, as we crept slowly down the road toward town.

The truck's muffler and exhaust system were long gone and the engine was so loud that we never attempted conversation. No sense competing with the sound of a working truck. I backed the loaded truck down onto the dock and was relieved to turn the ignition off while we yanked traps from the truck, lining the gear along the southern edge of the wharf. The opposite edge was lined with someone else's gear, waiting for high tide to facilitate loading the boats with relative ease.

"Dad, if you go back for a few more traps and some bait bags, I'll bring the boat in," I said as the last trap scraped across the open tailgate. The tide was still coming, but the depth of water beside the dock was ample at this point for the *Mattie Belle*, whose draft is 3 feet.

Dad cranked the truck up with a roar and headed back as I walked down the ramp to the dinghy float. Untying the painter and climbing into the 17-foot skiff, I thought how empty the float seemed, and how jammed it would soon become with the arrival of yachts and sailboaters who anchor out and come ashore in their motorized rubber inflatables. As the sound of my truck faded, I was feeling quite happy with the thought that I was now very close to having traps in the water, and that many good things would soon follow. My favorite summer residents would arrive, the blueberries would be plentiful, and the fishermen would all be ass-deep in lobster. A vision of dinner parties, my mother's lobster casserole, and blueberry pie had my mouth watering as I pulled the cord starting the 9.9-horsepower outboard motor on my skiff that would take me out to the *Mattie Belle*. Amazing, I thought, I'd had such good luck with this particular motor — it had been through so much but continued to run. The worst punishment it had endured had been last August at the hands of the Clemmer twins. Nathaniel and Tyler would soon be making their annual appearance, spreading total havoc for the two weeks their family spent here on vacation. The little hellions would be thirteen this year, I figured. As I motored across the thoroughfare to the *Mattie Belle*, I fondly remembered the twin boys, and realized I would have to learn to tell them "no" this August. Unlike Rita, they would not be avoided.

It had begun at a softball game. During the months of July and August, there is a nightly game in the field behind the Kennedys' house. I never missed a game as a kid and distinctly remember believing that either Danny MacDonald or Rob Dewitt might succeed in sending the ball into the weathervane on top of the church steeple with their turns at bat. There were often up to twenty players on each team, mostly kids from three to sixteen, and a few token adults to ensure some semblance of sportsmanship. One warm evening last August, taking my turn as token adult, I shared left field with the Clemmer twins.

The boys are identical in appearance, except for the presence of eyeglasses on the quiet one. The one without glasses is far from shy. "I'm not very good at softball.

Want to go fishing?" asked the nonquiet Clemmer as we patiently watched our eight-year-old pitcher sling the ball overhand, underhand, and sidearm in every direction but toward the plate.

"No. I want to play softball. When does school start?" I asked, changing the subject.

"I don't go to school. How's your outboard running?"

"Fine. You have to go to school. It's a law. Did you get kicked out or something?"

"No. I'm homeschooled. Can we go for a ride in your skiff?"

"Not now. We're playing softball. Are you going to be in Bernadine's talent show?"

"No. The rest of my family is though. Can we go fishing?"

"Not now. Why aren't you a part of the family's talent act?"

"Because I have no talent. This is boring . . . ball fifty-seven. Let's do something else," the not-shy Clemmer pleaded.

"Jesus! Are you a moron? We're playing softball! You don't like softball, you don't go to school, you have no talent. . . . I think you're a moron."

We laughed as our third pitcher bounced the ball off a spectator's head. "We'll never have a turn at bat," the not-shy twin complained. "How many outs do we have?"

"Zero." Before the half inning was over, our centerfielder had picked enough blue-berries for a cobbler, and the talkative twin had worn me down. I finally agreed to give him a ride in my skiff the following after-noon after hauling traps, if his parents gave him permission to go. We agreed to meet at the dinghy dock at four o'clock the next day, and that he would bring a life jacket.

Promptly at four the not-shy twin arrived with his quiet, bespectacled brother, pa-rental consent, and life jackets.

"Can you guys swim?" I asked.

"Of course we can! We're not morons," came the answer. "Can I drive?" Before we had pushed away from the float, the talker had just about worn me out with the con-stant "Why can't I drive? When can I drive? Can you teach me to drive? Can I start the motor?" I caved. Taking the middle seat be-side the quiet one, Chatterbox gave the cord a yank, and off we went. The boy maneu-vered the skiff smoothly and confidently around boats and moorings. He was a nat-ural. The quiet one needed a turn; it was only fair. Talkie reluctantly relinquished the driver's seat to his brother, who also proved himself a competent skiff skipper. While his brother carefully navigated the channel to Point Lookout, the talker once again drilled me with questions. "Can I borrow your skiff? Why not? I'll buy gas. Can I rent your skiff? I have seven dollars. When can I use the skiff alone?"

By the time we arrived back at the dinghy float, I had been adequately badgered, and agreed that, with their parents' blessing, my permission, and life jackets, the boys could use my skiff (sometime). After all, I was driving boats at their age. I made it clear that they were never to take the skiff without asking, because I might need it. Apparently I did not need it the following morning when the phone rang at seven A.M. "Can we take the skiff for a little ride?" I couldn't help but notice that "*your* skiff" had become "*the* skiff," and knew that before long it would be "*our* skiff." Knowing that "no" would never fly, I agreed and set the following guidelines:

1. Life jackets must be worn.
2. The motor must be kept at dead idle. No hot-rodding.
3. The skiff must be returned by ten A.M.
4. The skiff must stay north of the light house and south of Point Lookout.
5. The gas tank must not be left empty.
6. I must receive a phone call upon the skiff's safe return.

I had a million things to do. It was Sunday, and the state of Maine does not allow hauling lobster traps on Sundays between Memorial Day and Labor Day. So Sundays were great for catching up on paperwork and other shore-side projects neglected while hauling gear. I had just begun looking through the mound of bills on my desk when there was a tentative knock at the back door. Running down the spiral staircase, I opened the door to two very long twelve-year-old faces. It wasn't even eight-thirty yet. The quiet one had his hands jammed in his pockets and stared at his feet. The talkative one wasn't. "What happened?" I asked, knowing full well what had probably transpired to result in two sad, pale faces.

"Your motor just flew off the stern of your skiff. It's on the bottom." (I was back to sole ownership.) Now the two sad faces were joined by an irritated one.

"That's surprising. It just flew off while you were idling. You were idling, right?"

"Well, we were going to slow down."

"I'll bet you slowed down rather quickly when the motor bailed out. Come on. Let's go." Having learned what to do from my own misadventures with wayward outboard motors, I drove the kids to the dock and instructed them to pinpoint the overboard outboard's location while I tracked down my best fishing buddy, Dave Hiltz. The sooner the motor was recovered from the deep, the better the chances of reviving it. This was not the first time I had asked Dave to scuba dive for an outboard or to help me fix something. When I approach Dave Hiltz these days I never get a "Hello," but instead get a "What's broken?"

"Should I bolt this to the stern for you, Linda?" Dave asked as he placed the outboard in my truck after salvaging, flushing, rinsing, drying, lubricating, and starting it in his shop, using his tools while the talker watched and questioned his every move from draining to cranking. "Does he ever run out of questions?" Dave asked, smiling beneath his dark mustache. I thanked him profusely for his help (again) and climbed into the truck beside my young companions.

Before the Clemmer boy could ask to drive my truck to the dock, I said, "You know, I'm not mad at you. I'm sure it was an accident. And the motor is running, and no one was hurt. No harm done. Did you learn anything today?"

"Yes." His eyes sparkled for the first time since the knock on the door. "I've learned how to fix outboard motors. This whole thing is like falling off a horse. I should get right back on, right?"

"Not necessarily," I replied.

So today, I nudged the side of the *Mattie Belle* with the skiff and thought about the imminent return of the Clemmer twins. Like blackflies and mosquitoes, their arrival was inevitable. I knew that I would wind up letting them use my skiff. And I knew that I would once again have to rely on the kind salvage services of my friend Dave Hiltz.

I tied the *Mattie Belle* along the side of the town dock just as Dad returned with half a truckload of traps and a bunch of small mesh bags, which when filled with bait would be the size of grapefruits. My father began lowering traps onto the rail of the boat. As each trap came to rest on the rail, I grabbed it and stacked it in the stern of the *Mattie Belle*, and kept doing so until we had thirty-eight onboard, a comfortable load. Two bunches of ten buoys each were handed down to the deck from the dock, and I stowed them just inside the house's after bulkhead where I could easily reach them. Dad carried the bait bags aboard after leaving the truck in the parking lot; we let the stern line go and headed to Moxie.

The last chore to be completed on the road to setting out was to procure bait with which to fill the mesh bags or "pockets." The Island Lobstermen's Association stores heavily salted (pickled) herring on the tiny neighboring island of Moxie. Moxie is only an island at half-tide or better; low tide leaves a causeway between it and Kimball Island. Moxie is just big enough to house a bait storage building, a defunct outhouse that is used only in extreme emergencies, and a ramshackle shed in which lobster bands are stored. (Lobster bands are the stout elastics that are used to band lobsters' claws.)

High tide shortened the climb up the ladder on the end of the dock to just three rungs. As I climbed, Dad tossed an empty plastic "tote," or "fish box," onto the dock for me to pitchfork bait into. The old wooden door on the front of the bait storage building has large scales of peeling paint and a nasty message written across it by someone who was obviously pissed off at the world at the time of inscription. The door was held shut by a lobster trap that lay against it.

I was relieved to find three of the four fiberglass bait bins brimming over with freshly salted herring. Houseflies the size of bumblebees filled the shed with the sound of an electric fan. The handle of the pitchfork was gritty with dried salt and fish juice, and the nicks and scrapes my hands had suffered in the trap lot sung out in stinging harmony with the humming of the flies. Fresh bait does not stink, but it does have a musty odor.

I stuck the tines of the pitchfork into the top of the bin closest to the door, and dumped a fork-full of 8-inch blue-and-silver fish into the green plastic tote. I repeated the shoveling motion until the tote was full. On the back wall of the building, a small clipboard and a yellow lined pad of paper hung from a nail. A stubby pencil hung on a string beside the clipboard. The Lobstermen's Association believed in the honor system for bait sales. I penciled the date, my name, and "1 box" into the appropriate columns on the tally sheet and dragged my box of herring out of the building, closing the door with the trap.

We were glad to have fresh bait — anyone who has handled rotten herring knows to appreciate herring that is not. Island fishermen, at some point each season, come to appreciate bait of *any* quality, as we often find ourselves baitless. Our bait supply is delivered by a boat from the mainland, the owner of which is responsible for supplying Stonington-based co-ops as well. We Islanders have become increasingly aware of the fact that we are at the bottom of the bait chain. We get a bait delivery only when no one else on the planet needs any, and quite often are left with the dregs and scrapings from the very bottom of the boat.

I backed the *Mattie Belle* away from Moxie's dock and headed for the channel, the beginning of which is marked by a pair of navigational buoys maintained by the U.S. Coast Guard. As the boat split the distance between the two, Dad stuffed herring into bait bags. I ran the boat slowly, giving the diesel engine time to warm up to its normal operating temperature of 175 degrees. The needle of the temperature gauge rose to what I think of as "warm enough" as we passed Flake Island. I pushed the throttle ahead until the tachometer indicated that the engine was turning at 1600 revolutions per minute, just ahead of the vibration range of 1500 RPMs that shakes my radar slightly in its bracket. The upper right-hand corner of the chart plotter displayed a series of numbers indicating a course made good of 290 degrees, at a speed of 12 knots.

The *Mattie Belle*'s top speed is 23 knots, but I rarely run her that hard, due mostly to the cost of fuel. Horsepower is nothing more than fuel and air; the more of each you can force through an engine, the more power and speed. Well, maybe it's not quite that simple. There are the variables such as reduction gear, propeller size and pitch, and the shape and length of the hull. Andrew Gove had for a long stretch the distinction of owning the fastest lobster boat in Maine. His boat was clocked at nearly 50 miles an hour, leaving the competition in the dust — or spray, I guess. Andrew and Rose's home is full of trophies and awards won at lobster boat races sponsored in various ports up and down the coast. It is said of his boat that she'll pass anything but a fuel pump. The *Mattie Belle*, when she was built in 1986, was one of the fastest boats around, but there has been a trend in recent years to power lobster boats with engines of much greater horsepower than ever thought possible. My boat is by no means a dog, but I believe Gove's boat goes faster in reverse than the *Mattie Belle* goes forward.

Between the oil pressure gauge and the engine temperature gauge on the panel mounted on the overhead, the ammeter showed a slight tilt right of center, evidence that the boat's alternator was indeed functioning properly, keeping the batteries charged to just shy of 14 volts. All electricity on the *Mattie Belle* is DC, and all systems are 12 volts. I had learned the hard way to keep an eye on all of the instruments in a boat's panel, and not to rely on auditory alarms. Salt water dislikes everything. I once trusted an alarm that failed to signal low oil pressure, resulting in an explosion, a fire, a tow to a boatyard, and a hefty bill to rebuild the fried engine. More recently I had neglected the ammeter when the alternator had thrown its belt, and ran the batteries down to totally flat. Mistakes are expensive. Mistakes are good, because we can learn from them. I must be a slow learner because I repeat most of mine.

I headed the *Mattie Belle* toward a starting point off the southwest corner of Merchant's Island. It would be a relief to have this first load of gear sitting on the ocean floor, where the traps would finally, after all this handling, begin to fill with lobsters. Soon we would have completed one full cycle of setting out. Within two weeks of setting a load every day, we would be done setting, and the fishing would begin. I thought ahead to our first hauling day and hoped it would produce enough lobster to begin to pay for the fuel and bait. I hadn't looked at the balance of the boat account's checkbook since last winter, and I hoped we would not be operating in the red while waiting for the lobsters.

All thoughts of finances left me as we neared our waypoint and the depth sounder rose to show 10 fathoms, 60 feet, of water under the boat. This load of gear was rigged to fish a range of depth from 30 to 60 feet. Perfect. I pulled the throttle back to idle, and the gearshift back to neutral, and maneuvered the boat so that we would drift, leading with her starboard side, the side we worked from. This ensured that I would see any gear we might drift onto, and before doing so could bump the boat in and out of gear to avoid cutting line with the propeller. As the gear starts to thicken in an area, avoiding line becomes a real challenge.

Working our way in an easterly direction along the south shore of Merchants, we set pair after pair of traps. The stack of traps got smaller and smaller and was finally gone. As I steamed toward home, Dad hosed off the deck where the traps had left pine needles and dried mud from the trap lot. Dad and I had been spending a lot of time together. It was nice. We had been close when I was young. I was the ultimate tomboy, preferring hunting and fishing and boating with Dad to anything my friends might have going on, right up until leaving for college and the extended (seventeen-year) fishing trip. I was glad to learn that some things had not changed.

My back ached, but I didn't dare complain to my seventy-one-year-old father, who had done most of the lifting. Our backs would be much worse by the time we had set the entire bunch of gear, but anticipation of another banner lobster season would get us through aches and pains, mosquito bites, sunburns, even the Clemmer twins. My mother would have a wonderful dinner for us tonight, as she always did. And she would carry more than her share of the conversation, entertaining Dad and me with the most delightful stories about anything or nothing. She would be animated, dramatic, and amusing, as my father remained the straight man. It was fun. We would eat, drink, and laugh with the lighthouse as a backdrop until the curtain of darkness fell, cloaking all but the red flash from Robinson Point Lighthouse.

THE LIGHTHOUSE COMMITTEE

Against my better judgment, I became a charter member of the Island Lighthouse Committee. It seemed, at the time, a way to perform some sort of civic duty without getting involved in town politics, which have a history of being bitterly divisive. Even the most insignificant article on the agenda for an official meeting of any of the town's boards (school, ICDC, selectmen) has the power to divide the community. Town meetings are often conducted through verbal assaults launched by someone from one side of the town hall against a neighbor on the other. A few folks find this entertaining. The majority is sickened by the lack of civility. As much talk as there is about "healing the community" and the disgust we feel in the pits of our stomachs when another scathing letter is posted for public display, the rifts just keep getting wider and deeper. The best shot residents have at attending a meeting that won't necessarily keep them awake at night is to ensure that Ted Hoskins is on-Island before venturing to the town hall to cast a vote.

Ted Hoskins is many things, but primarily he is the minister of the Island's Congregational church, the only place of formal worship on the rock. There is something about Ted that separates him from the rest of us, yet allows him to be an integral part of the whole. I believe this is only partly attributable to his ministerial profession. It is mostly due to some personal quality that quietly shies from description. And to me this makes him a bit mysterious, in an interesting way. Recently, I confessed to two female friends that I have a secret crush on Ted Hoskins, to which they both proclaimed, "So do I!" I suspect there are few men or women of any age who do not harbor special feelings for Ted. So perhaps "crush" is the wrong word.

When I imagine God in my mind's eye, I see Ted Hoskins without his glasses. (Why would God need glasses?) Ted enjoys catching fish, which is what I like most about him. It's comforting to have fishing in common with someone I can think of as a link between me and things I don't understand. It's like having a buffer zone. A large man with snow-white hair and beard, Ted is an impressive figure in any company. His voice is my other favorite thing about him. I once stated that the sound of Ted's voice was nearly enough to tempt me to attend his Sunday services. His soft, yet deep purling cadence is so delightful, even in casual conversation, that although the message might not always mesh with my peculiar beliefs, I think that I might enjoy a sermon of his simply for the sound, as though it were a concert. Attending a formal church service solely to hear the minister's voice, though, hasn't been reason enough to get me to do more than contemplate climbing the long, steep boardwalk on any given Sunday morning. The uphill hike to the church, the steeple of which can be seen above the highest greens from quite a distance, is a bit of a pilgrimage. So when I see Ted he's not at the pulpit but bent under the hood or over the engine of a car or boat, whichever he is tinkering with at the time. And of course, I am always relieved to see Ted at a town meeting, where he will most assuredly be voted in as moderator.

Ted's presence tends to encourage people to keep the language cleaner and behave in a more civil fashion. This, in turn, keeps meetings moving forward. I have attended meetings when Ted did not, where the first item up for discussion becomes like a car stuck in a ditch. Some passengers are helpful, getting out to push, while others, fearing exposure, stay huddled within the vehicle. The driver, whose finesse wanes with waxing frustration, puts the pedal to the floor, miring the car farther in the ditch — then abandons both car and passengers and walks home. Neither passengers nor pushers are interested in salvage rights, so there the car is left. The driver, now a pedestrian, fumes, then pouts, and eventually blames not the ditch but those who were anxious to accept a ride yet were hesitant to push. Nasty letters are written and people stop talking. The godliest role Ted plays on the Island is that of mediator. With Ted around, fewer vehicles tend to hit the ditch.

I never thought mediation would be needed between the Lighthouse Committee and the town fathers. This committee, I thought, would be one the entire Island community could rally around. I foresaw nothing even remotely controversial when I agreed to become a member of the newly formed group. Perhaps I should have followed Wayne Barter when he resigned as a member after just one meeting, for Wayne avoids conflict as assiduously as I do. Politically naive, I thought the Lighthouse Committee would be recognized for doing something good for the town. I imagined the project would be unanimously embraced. Being stamped with the committee title of "community liaison" should have been a red flag. When asked to assume this role, rather than "Sure," I should have asked "Why?" But you know what they say about hindsight.

The horror show began on a fine day while I was walking home. At the foot of the hill that terminates in Billy and Bernadine Barter's yard, I navigated a drainage ditch to avoid running into Jeff Burke's old Jeep Wagoneer. The Jeep was parked as close as Jeff could manage to Bernadine's entryway to minimize the steps involved in loading clean laundry onto, or unloading dirty laundry from, the luggage rack on top of the vehicle. Bernadine, who never has any trouble keeping busy, is in the laundry business during the summer months, her only client being the Keeper's House. The Island's only bed-and-breakfast, the Keeper's House was originally the home of the Holbrooks, the first family employed as keepers of the lighthouse tower back in 1907.

Banana boxes now full of clean white sheets, towels, and tablecloths were neatly stacked on the roof of the Jeep, and awaited delivery by Jeff to his wife, Judi. Jeff and Judi together own and run the tidy business on Robinson's Point. The Keeper's House Inn is, next to lobstering, the largest employer of Islanders. The Burkes offer several jobs cooking, washing dishes, waiting tables, and cleaning rooms at the Inn. Although the nightly rates for rooms are considerable, the Inn is booked solid months in advance, the majority of the rooms occupied by lighthouse lovers from God-knows-where. Jeff and Judi like to refer to their clientele as "lighthouse aficionados," but I prefer "freaks" or "fanatics." The true freaks are highly recognizable and nearly always female, sporting lighthouse clothes, handbags, and jewelry. Any woman from whose earlobes swing lighthouse towers or from whose arm dangles a lighthouse purse has got to be a guest at the Keeper's House. The men are less obvious, and unless accompanied by a woman whose jacket displays pictures of every lighthouse on the eastern seaboard, they can go about the business of lighthouse obsession undetected.

As of 2002, a single night at the Keeper's House was running somewhere in the range of $250 to $300. This charge includes three meals, two of them served at the house. For lunch, each guest is handed a bag after breakfast and politely told to "get lost" until dinner. A few of the guests have done just that, disappearing on foot or dilapidated bicycle. In fact, I recalled, as I stepped from ditch to driveway behind the Jeep, that the last time I saw Jeff, he was searching for two of his aficionados who had not shown up for the evening meal.

"Hi, Linda," Jeff called to me as he closed the Barters' door behind him.

"Hi. Beautiful day." I smiled and kept walking along the rutted and potholed road that would eventually end at my folks' house.

"Sure is. Say, if you have a minute, there's something I would like to discuss with you." And the minute that I did have was my introduction to Jeff-the-salesman. The man is good. He knew exactly what to say. A perfect approach and a short sales pitch was what had me enthusiastically joining the new Lighthouse Committee. It isn't that Jeff's voice is effective — he's no Ted Hoskins. Jeff's voice is flat, lacking inflection, and he sounds as if he's whining even when he isn't. But the text is so effective that the delivery becomes inconsequential. Jeff's pitch, which somehow left me feeling as though I'd received a very privileged invitation, boiled down to this: The federal government was now authorized through an act of Congress to transfer legal ownership of thirty-five Maine lighthouses from the Coast Guard to "eligible entities." The proper maintenance of lighthouses had been shortchanged in federal funding; many of the towers were in varying states of disrepair. The Maine Lights Bill established the Maine Lighthouse Selection Committee, whose charge was to work together with another organization, The Island Institute, to identify eligible applicants for ownership, review proposals, and approve one applicant for each of the thirty-five lighthouses. The requirement was that eligible entities needed to exhibit an ability to restore and maintain these structures; if they didn't, ownership would not be transferred. The implication was that unwanted towers would probably be torn down. Entities considered eligible would be given preference in the following order of priority: federal agencies, state entities, local governmental entities, nonprofit organizations, educational agencies, and community development organizations.

The Island, as an organized town entity, would fall in the middle of the priority pecking order, and would be the most logical recipient of Robinson Point Lighthouse, which happened to be one of the thirty-five up for grabs. Neither Jeff nor I could imagine any other interested party. And the townspeople certainly could not bear to see such a historical landmark and point of heritage and pride be torn down or allowed to disintegrate. Jeff thought I might be interested in getting involved because of my family's ties to the lighthouse. He was right.

In 1906 my great-grandfather Charles Robinson sold a two-acre parcel of land at the tip of Robinson's Point to the U.S. Government for the construction of a light station. Maine legislators finally succeeded in pushing the bill through Congress after thirteen years and sixteen legislative sessions. The Robinson Point Project was the last of many commissioned by the U.S. Coast Guard, and included the construction of a 48-foot tower, Victorian-style lightkeeper's house, a wood shed, a tiny oil-storage house, and a privy.

On Christmas Eve of 1907, eight-year-old Esther Holbrook lit the lantern's wick for the first time. The wick burned for twenty-seven years, guiding countless fishing vessels safely into the passage leading to the town's anchorage. It burned through two families of lighthouse keepers, until the financial problems of the Great Depression forced the U.S. government to trim its maintenance out of the budget. The Robinson Point tower had by then been automated, and so would be owned and maintained by the Coast Guard. But maintenance of the keeper's dwelling and outbuildings, which were no longer needed, was not funded. The light station, along with eight others along the Maine coast, was auctioned off in 1935 by the government. With the help of Senator Margaret Chase Smith, my grandparents were able to buy back all but the tower.

The keeper's house became my family's summer home. It's where we went when we weren't in school, and where we longed to be when we were. Not many people can say they have lived in a lighthouse. The times we spent in the house, my siblings all agree, were magical. We slept at night in rooms whose walls were massaged every four seconds by the soft, red wink of the tower. The house seemed to have a life of its own, protecting us and the memory of those who had lived there before. I've never felt so at home in any other dwelling, and perhaps never will. There are those who believe that the Keeper's House is haunted. I think spirited is more accurate.

Three generations of Robinsons and descendants, mostly Greenlaws and Bowens, enjoyed the light station for the next fifty years, until 1986. Joint ownership rarely works. The five children of Aubrey and Mattie Belle Robinson Greenlaw shared ownership and tax burden equally, but use of the house was never equitable — my immediate family getting the lion's share of the summer months. In a "majority rules" decision, the five offspring of Mattie and Aub sold the lighthouse property outside the family, believing that transaction to be the only fair solution. The lone dissenter, Uncle George, has nursed a grudge for nearly sixteen years.

I now stood in the road talking with the present owner of what had become a thriving business, the Keeper's House. I have the privilege of admiring the light from afar, and the pleasure of fond memories. The tower is the bulk of the view from my parents' place, and I dreamt that sometime in the future it could be a focal point of the windows in a house of my own. Past, present, and future all came into play in a quick decision to join a committee whose goal would be to obtain, restore, and maintain the tower for and with the town.

I left Jeff with his banana box pyramid of immaculate linens after agreeing to attend a meeting that night at seven. I felt I was doing good, and was anxious to share the news of my involvement with my mother, who sat engrossed in a book when I barged through the kitchen door. My mother, who has always had more than her share of energy, even reads enthusiastically. Someone who thoroughly enjoys literature, Mom laughs out loud, sighs, and cries her way through a good book. Sitting quietly and relaxing with a book is not in her repertoire. She is constantly tapping a foot or shaking her head. She doesn't just turn pages, she whips them aside from right to left, nearly ripping each from its binding.

I don't know if there is such a thing as a "reading addiction." But Mom gets nervous when she gets low on reading material. Years ago I mentioned to my mother that two of her bathrooms were out of toilet paper, and the third was on its last roll, to which she replied, "I'm going out to buy books on Friday. I'll pick some up." The older she gets, the worse it is. She starts warning my father (or anyone else who will listen) days in advance of a book shortage: Jim, I'm down to two books. I'm on my last book, Jim. I'm going off on the early boat — I need books. I thought I had another book here — have you seen my book? And off she goes for a fix, savoring the final pages, which she has hoarded to sustain her for the thirty-minute boat ride. If she were not such a poor driver to begin with, my mother would be one of those idiots who read while operating a motor vehicle. I'm not complaining, but I think I've been raised by a reading addict.

My mother had four children over such a long span of years that first menstrual cycles and diaper rash were often dealt with on the same day. My mother's theories and practices in child rearing created an interesting household. Mom was not unduly strict; in fact, we got away with a lot. My mom's mother (bless her soul) explained to her daughter when I was a baby that there exist some children who simply cannot be spanked, and that I was one of them. How fortunate! There were things, however few, that my mother would not, under any circumstances, tolerate in her children. We were forbidden to chew gum, use the word *hate,* or tell someone to "shut up." ("Shit"

and "goddamn" were OK, even from tod-
dlers. And "frig" was the root of all verbal
communication: frig, friggin', frigger, royal
frig. I even had a T-shirt that said, "Friggate
is not a dirty word in Bath, Maine.") Most
important, and key to my mother's child
rearing, was a very limited tolerance for
whining.

My mother adhered to a strict policy: She
would only hear complaints or kiss boo-
boos after she declared "the clinic is open,"
which usually occurred at precisely four P.M.
daily. Any sniveling prior to four would be
cut off with a sharp "the clinic is closed" or
"the doctor is not in." At four o'clock we
would line up and wait for our turn with
"the doctor," who would dispense anything
from Midol to baby aspirin. Four o'clock
was also the time for pep talks, consultation,
and consolation. My mother was and still is
the best ego-booster and self-esteem builder
on the planet. Even when I had no physical
ailments, I would visit the clinic and relish
Mom's time alone with me. Not a day
passed without my mother making me feel
special in some way. You might bleed to
death or cry your heart out while waiting for
the clinic to open, but if you could survive
until four o'clock, there wasn't anything my
mother couldn't cure.

"Oh, I think it's wonderful that you're getting involved in something other than lobster traps. Who are the other members of the committee?" Mom sounded pleased and interested when she looked up from her book.

"Well, Jeff Burke is the head of the whole thing. Let's see . . . he mentioned Dave Hiltz, Dave Quinby, Elaine Bridges, and Wayne Barter."

"Jeff Burke? Heading up the Lighthouse Committee?"

"Yeah. Why are you looking at me like that?"

"Jeff Burke is forming a committee to get ownership of the lighthouse tower for the town?"

"Yes."

"Jeff Burke?"

"Yes, Mom. Jeff Burke. You know, the guy who owns the Keeper's House?" I quipped sarcastically.

"Precisely."

And that was the beginning of the first battle with my mother. After having been away from home for the better part of the past seventeen years, it was only then that I realized our mother/daughter relationship needed redefining. When I first began swordfishing I was nineteen, barely an adult. Now, at middle age, I had returned to the nest. A child needs a mother. I had a good one. As an adult, I suppose I still have the need to be mothered, but relationships between women are complicated and confusing. I had a lot to learn. In all my time at sea, I had developed no friendships with women. I had little or no contact with women. I worked for, with, and around men. My bosses, shipmates, crewmembers, and fishing buddies were all male. Once I returned to the Island, my parents became my closest friends. In the spirit of "only a true friend would tell you this," my mother often said things that I did not want to hear. And in the spirit of "mother knows best," she was usually right.

The problems inherent in Jeff Burke's leadership of the Lighthouse Committee were obvious to most everyone once some time had passed. My mother recognized the conflict before we had gathered for our first meeting. I thought she was nuts, and I may even have voiced that. I never, until now, admitted that she was right, nor did she ever say "I told you so."

"Let me get this right, Linda. Jeff the philanthropist is unselfishly taking on this project for the people of the town?"

"Yes. It's going to be a huge amount of

work. You should see the application process. We'll probably need to raise $100,000 to restore the tower."

"And Jeff is willing to commit that much time and effort for the good of the town?"

"Well, someone has to do it. Why not Jeff?"

"Does a $100,000 lawn ornament mean anything to you? What kind of business would the Keeper's House do with nothing to keep?"

"Oh, for Christ's sake, Mom . . ."

Oddly enough, our debate began at four that afternoon, and many subsequent ones would end with the pouring of the first scotch and water. It was the clinic schedule all over again. My mother agreed that the town should own, restore, and maintain the lighthouse tower. Her problem with the project was Jeff. "I like Jeff, but I question his altruism in this case" was her platform.

Many of the older residents of the Island, my folks among them, consider the Burkes hippies and rabble-rousers. Jeff and Judi are activists who are committed enough to social and political causes to travel to third-world countries in order to feed the hungry or join whatever protest is percolating at the time. When I see Jeff assume his posture of purposeful wanderer, armed with sheets of paper, I know that he will approach me to sign a petition. Jeff's petitions run the gamut from town business to worldwide reduction in nuclear arms. If the lighthouse project was another of Jeff's causes, and even if his motives were selfish, I couldn't have cared less.

Organizing and conducting committee meetings, writing the application, preparing for special town meetings, and fund-raising would be an immense amount of work; most people didn't appreciate this. I doubted that anyone without some motivation from potential personal reward would have taken on such a task. Other people felt differently. Mom and I went several rounds debating motivation, inspiration, and philanthropy, and all the while the Lighthouse Committee was building steam with Jeff shoveling most of the coal.

The first order of business for the committee was to inform and educate the citizens of the Island about the Maine Lights Bill, with hopes that a majority of voters would authorize the town's board of selectmen to apply for ownership of the tower in the name of the town. This authorization passed the town's scrutiny overwhelmingly, 41–2, at the first special town meeting held at the Lighthouse Committee's request.

The only apprehension — that property taxes or other resources of the town's coffers might need to be used in the project — was quickly assuaged with the promise that all funding would be from private donations and grants. The Lighthouse Committee would have to raise the money for restoration and maintenance of the tower with no help from the town's budget. "Property taxes are high enough." Fair enough. All members of the committee were confident that the money could easily be raised.

People tend to get emotional and defensive when money is discussed. I let it be known right away that I cannot bear the thought of approaching anyone with a request or plea for money, and I would not be an asset on the fund-raising team. I was glad not to be in a position where I was expected to ask anyone for a donation. The sentiment among some of the more affluent summer residents, who are constantly hit up for financial support for Island projects and causes such as school field trips, was that Jeff should lead the charge with a major contribution. The opinion was voiced on several occasions to the community liaison (me) that the Burkes would be the only ones to gain financially from the project, so Jeff should "put his money where his mouth is."

The Burkes were willing to approach their clientele for contributions, but felt that pledging a percentage of the Inn's proceeds, as suggested by many skeptics, was beyond reasonable.

Before any funds could be raised, regardless of from where or whom, we had to be approved as worthy guardians of the tower through the application process. The selectmen and committee members would have to work quickly and closely to meet certain deadlines, such as the drop-dead date for submission of the application to the Maine Lights Program. Problem? Jeff Burke and first selectman Matthew Skolnikoff were archenemies. These two men are not fickle in their dislikes, and their hatred for each other has withstood the test of time. Both are men of intelligence. Both are exceptional and prolific writers. The townspeople were now treated to eloquent and passionate letters of attack and defense. Some letters were drafted under the guise of informing concerned citizens. Some of the bashing appeared, editorial style, in our local newspaper. None of the public correspondence from either came right out and said "I fuckin' hate you," although most everyone understood that was the message.

The seed of discontent had been planted

some time ago when Matthew's employment at the Keeper's House ended abruptly. I never knew whether he had quit or been fired, but that seed sprouted a sapling that now loomed over the lighthouse project. Eventually, someone was able to convince Jeff that he needed to remove himself from the committee, which he understood was the only way to get anything accomplished, and he did so willingly enough. Jeff now attended meetings as a concerned member of the community. Oh, he still did 90 percent of the committee's work and still ran the meetings, but someone else would now deal directly with the selectmen. This was much better. The Keeper's House was adequately represented and Elaine Bridges became our (figurehead) chairman. Elaine was employed at the Inn as a cook, as was committee member Dave Hiltz's wife. Lisa Turner, who joined the Lighthouse Committee midway into the project, when fresh blood and enthusiasm were needed, was also one of the Burkes' employees. Dave Quinby derived none of his income from the Keeper's House, but considers Jeff his best friend.

Before Jeff resigned from official member status, he did some work to enlist volunteers for fund-raising. Had Jeff been aware of the "clinic discussions" that had been taking place nightly, he never would have knocked on my mother's door in an attempt to convince her to help with raising money. Unfortunately, I was an uncomfortable witness to a very polite conversation in which my mother raised pointed and direct questions about Jeff's motives. She told Jeff, in a nice way, that he was not trusted by everyone in the community. I wanted to evaporate. Jeff explained that he understood how some people might assume that he was acting solely with an interest toward his financial security, but that simply was not the case. The last thing Jeff said, in what had turned out to be more of a session of self-defense than a call for volunteers, was "my intentions are noble." My mother walked Jeff to the door, wished him a good day, and vowed to give fund-raising some thought. But I knew she wouldn't.

Some of the Island's residents became confused when more special town meetings were called to confirm, or revote on what had passed by an overwhelming majority in the original vote: to authorize the selectmen to apply for the title of the tower. "Didn't we vote on this before?" echoed through the town hall. A phenomenon of small-town politics is the ability of the "two opposed" to not give up the ship to the "forty-one in favor." It may appear that the exact article is passed three or four times, but those with political savvy can point out the subtle differences that make it legal to bring it up again and again in hopes that people's sentiments have flip-flopped since the last meeting. Sometimes revotes are justified in terms of a technicality, as with a mistrial. Sometimes revotes are justified due to some new bit of information unknown at the time of the preceding affirmation. Mostly they are just demanded.

Residents began to get weary of attending meetings about the lighthouse. Many were sick of casting votes. Most were bored with the latest bout of Matthew vs. Jeff. On the Island, it is never enough simply to exhaust a topic; it needs to be replaced by something or it will go on being discussed no matter how little more can be said. Someone would have to do something stupid or gossip-worthy soon; the lighthouse controversy had worn thin. I craved conversation that did not include what had become the saga of the light. I prayed for distraction.

My prayers were answered when "the Alabama Slammer" arrived on Island. She certainly wasted no time becoming the talk of the town, giving the Lighthouse Committee members time to catch our breath. Her name was Suzanne, and she was an old friend of Victor Richards, one of our most unusual local lobstermen. A sixty-five-year-old ex-crop-dusting pilot, Victor has a unique nicotine habit. He eats cigars. He never lights them, doesn't even spit, just gnaws until the stogy disappears. Victor's old job took him all over the country. At one time he found himself employed in Alabama, which is where he became acquainted with Suzanne.

Victor is crazy about women and Harley-Davidsons. So a woman who rides a Harley is of particular interest to Vic. Suzanne, with her long red hair and Southern accent, is the quintessential biker chick, although I don't recall any tattoos. The residents of the Island had never seen anything quite like her. The girl couldn't keep her shirt down. I quickly heard that she had bared her breasts to Cal Lawson, captain of the mailboat, on her way to the Island, and upon stepping onto the town dock gave some of the local fishermen a peek, too. I assume that some of the reports of flashings were exaggerated, due to the shock of it all. That's not to say that the locals are prudish. But, blame it on the weather, we Maine gals tend to keep our clothes on. And when a fisherman refers to someone's language as "unfit," it probably is.

My initial introduction to Alabama was the evening of Victor's second Ladies' Night Out, an event he sponsored in order to get women into his house and ply them with alcohol. Victor's first of such parties had been quite a success. I think there were seven females and Victor — good odds for Vic. We drank, laughed, and watched videos of Russian mail-order brides. Prior to Suzanne's arrival, Victor had seriously contemplated a trip to the former Soviet Union to pick out and purchase a mate for himself. I tried to convince him, as we rated the video women on a ten-point scale, that the candidates would not find life on the Island terribly appealing. I could imagine one of these eighteen-year-old Russian beauties wishing she were back in the USSR after ten minutes with a grouchy senior citizen with tobacco stuck in his teeth, living on this rock with no Bloomingdale's. "Vic, I think these girls are dreaming about moving to New York, or L.A. If you buy one and bring her here, she might be unhappy or disappointed."

"Disappointed? With me? No way!"

"No, not with you. Hell, no. Jesus, what was I thinking? Hey, this one looks good. I'll give Natasha a nine."

So Suzanne's reappearance into Vic's life saved him both a trip to Russia and the rubles determined to be fair market value for a wife. There would be no bride-shopping at Vic's second party, which I had no intention of attending. I had my nephew, Drew, who was ten at the time, and his friend Trent, for an overnight at Aunt Gracie's. The boys had plans to consume junk food and horror movies. When the phone rang five minutes into the first scene involving a teenager and a butcher knife, I was more than happy for the interruption.

She didn't need to tell me who she was, although I had not yet met her. I could tell from the accent, music in the background, and noise in general that it was Suzanne calling from Victor's party. She explained that she wanted to meet me, and that I should come over and join the festivities. I explained that I had two ten-year-old boys for the night, to which she said something terribly rude. Even more reason not to go to Vic's, I thought, and prepared to hang up the phone, thanking her for calling. Alabama was not taking "no" for an answer and swore she would come over and get me if I did not make an appearance within the next five minutes. When she added that she had nothing on but a pair of cowboy boots, I realized that her boots were made for walking, and that she had every intention of letting them take her drunken nakedness to my front door, which was only about 200 yards from Vic's. (Talk about horror movies . . .) I feared the boys' early education, and left the kids glued to the TV, vowing to return in fifteen minutes, which ended up being closer to ten.

Merle Haggard was growling about one of the more miserable portions of his life when I entered Vic's house. The kitchen was a bit smoky, but clear enough for me to see that everyone was fully clothed. So Suzanne had been bluffing, which was a relief. My cousin Diane sat at the kitchen table, where she leafed leisurely through a *Down East* magazine. Merle managed to overpower several conversations in the adjoining room, where Victor appeared to be enjoying the company of half a dozen women. "Hey, what are you doing in here alone?" I asked my cousin.

"Oh, you'll figure it out," Diane smiled, raising a rum and Coke to teeth that are whiter than teeth need to be.

I shrugged and grabbed a beer from Victor's refrigerator. One step over the threshold between kitchen and party zone, I was greeted by the very attractive redhead, who I knew must be Alabama. "Well, it's about fuckin' time," she yelled.

"Well, thanks for inviting me." I grinned at Victor, who seemed quite proud of his friend.

When I turned my attention back to Suzanne, she snatched the front of her shirt up over her face, exposing boobs as white as my cousin's teeth, and shrieked, "These are Alabama tits!"

I don't know what came over me. I couldn't help myself. I waited for Alabama to tug her shirt back into place so that I had her full attention. I pulled the top of my T-shirt away from my collarbone, and looked down through the extended material at my own breasts. After a long, thoughtful inspection, I released the neck of my shirt, which snapped back against my skin, shook my head at Suzanne, and advised, "If that's the best Alabama has to offer, you should really keep them covered." Diane broke into a gale of laughter, and I later learned that she had done and said the same thing when confronted with Alabama tits earlier that evening. (I guess Greenlaw women are hard to impress.) I said good night to all present and, taking my beer with me, returned to the better-behaved partiers at Aunt Gracie's.

The breast occurrence was a minor incident compared to other activities that were reported to have transpired at Vic's after my departure. Stories spread like spilled paint over the Island the next day. And like most stories, they improved with every telling.

The single night at Vic's was the root of all gossip for at least two weeks. No one was discussing the lighthouse. It was great. The Lighthouse Committee's dealings with the selectmen were quite pleasant during this period. The project was progressing nicely. The application was written and sent out before the deadline, and as far as we knew, no other entity had any interest in our tower. We were now confident enough to start raising funds for the restoration. Jeff had succeeded in obtaining some very enthusiastic help from a number of talented Islanders; summer residents and year-rounders alike pitched in. We had put the job of restoration out for bids and had received several. The fund-raising target number was $75,000. Suzanne's exploits were still being discussed.

A letter was drafted and sent to hundreds of potential contributors. And the town continued to be absorbed in Vic's new friend. Grants were researched and applied for. The pet topic of conversation was still Alabama. Tax-deductible dollars were pouring in. The

Lighthouse Fund was growing quicker than even the most optimistic had anticipated. Suzanne was still the most interesting tittle-tattle about town. It was like a sleight-of-hand trick by a magician. While all eyes were on Suzanne, waiting for some performance, the committee was really getting important things done. No one seemed to care about us anymore. I knew this couldn't last forever. Surely someone would commit some act of stupidity to take the focus away from Suzanne. I would never have guessed that a member of the committee itself would be the one to supply the tidbit worthy of wagging tongues not yet exhausted from saying "bare chest." Soon enough all the negative scuttlebutt the Island had to offer would come right back to the tower project.

To Suzanne's credit, after Vic's party, she didn't appear to drink at all and had been working in the stern of Vic's lobster boat every day. She was keeping her clothes on in public, and everyone who got to know her liked her. But she got no relief from the rumor limelight until Elaine Bridges, the chairman of our committee, took center stage, which resulted in the brakes being applied quite abruptly to the success and progress we had enjoyed while no one was paying attention.

★ ★ ★

A group of fishermen leaned against the sides of the bed of Dave Hiltz's pickup truck. It was late afternoon, and the group showed signs of wear from the long day of lobstering. Dave had driven his truck down onto the end of the town landing to load his family's portion of the co-op order that had just arrived by boat; the order consisted mostly of fruits and vegetables. (Dave seemed relieved that my approach was not with a request for a favor.) It's interesting that men from whom a "hello" in passing on the road is the entire conversation, when gathered around a tailgate, can talk for hours. A truck is the fishermen's roundtable. The discussion across boxes overflowing with salad ingredients was just as fresh as the produce. It was pleasant talk among fishermen who were happy to be tired. The lobsters had not yet begun to crawl out of hiding, so the day's harvest had not reached any degree of profitability. The discussion was, for the most part, a continuance of a long stretch of wondering "if the damn things would ever shed." I knew I had to trudge my weary legs up the hill and into my own truck, but leaning on Dave's was so comfortable, I was hesitant to move. It appeared that the other leaners felt the same way. There was talk of hot showers, cold dinners, and sleep, but no one made a move.

Judi Burke appeared in the Jeep Wagoneer, which sports a model lighthouse just in front of the luggage rack on its roof. It is always odd to see Judi during the frenzy of the Inn's busiest season; she seldom escapes the responsibilities incumbent in bed-and-breakfast ownership. I see Judi in town no more than once or twice a month between May Day and Halloween, the period the Keeper's House is open for business. I've observed through the years that Judi's physical appearance mirrors the Island's seasonal changes, and that if there existed no monthly calendar, I would need only to see Judi to know whether the day was in April or November. This afternoon was in mid-July, but as she approached I saw late October in Judi's face. Although her greeting to the group was friendly as always, there was clearly something wrong. Normally, Judi's energy doesn't flag until the days shorten up and she is anticipating the departure of the last wave of guests at the Inn. When Judi asked to speak with Dave and me alone, I knew that I had mistaken deep concern for fatigue. Judi wasn't tired; she was worried.

Dave and I were informed that there was

some problem having to do with the lighthouse, and that we were expected at an emergency meeting of the committee at the Burkes' that night. We thanked Judi for delivering the message. Judi left. Although she had been careful not to divulge any details on the nature of the problem that had suddenly arisen, Dave and I sensed that tonight would be about damage control.

Dave shook his head as he climbed behind the wheel of his truck. His mustache showed no hint of his usual smile. "I don't know what this meeting is about, but I can tell it's not good."

In fact, it wasn't good at all. Seven o'clock found the members of the Lighthouse Committee staring at one another across a room full of near despair, waiting uncomfortably in the most uneasy atmosphere I had ever experienced. I tapped a foot nervously; Dave Quinby fidgeted; Elaine Bridges, our chairman, cried. It is difficult to watch someone sob inconsolably. I fought the urge to cry myself. I'm sure all present were thoroughly anguished by the time Judi, who along with Jeff had tried unsuccessfully to soothe the distraught young woman, said, "Elaine has something to tell you."

I was relieved that someone had broken the silence and now waited anxiously for

Elaine to pull herself together enough to speak. I wondered how many in the room understood what it's like to cry so long and hard that speaking is impossible. Words come out, two or three at a time, then the next syllable catches in your chest, causing something close to a hiccup. These hiccups are deep, and cause the crier to shudder, resulting in a mixture of ache and nausea — a series of short, jagged breaths allow the blurting out of the next couple of words with a steep stop between each individual sound. This was not the crying of anger, mourning, or pain. This was the crying of shame.

All I managed to understand was that Elaine was apologizing to us for something she had done, some mistake. Then she was again wracked with sobs. Dave Quinby comforted and encouraged Elaine with words that expressed what we all felt. He said that we were friends, and that in order for us to help we needed to know what she had done that was causing her such anguish. This proved to be exactly the right thing to say. Elaine spilled her guts.

We quickly learned that Elaine's "mistake" was that she had "borrowed" some money from the Lighthouse Fund for the purpose of paying personal bills. The mistake, which was actually three or four separate errors, would soon be discovered, as the town's books had been sent for their annual audit. Elaine, who was our chairman, secretary, and treasurer, had the sole control of, and access to, the money that had accumulated from donations for the lighthouse project. She had written a series of checks for relatively small amounts to herself and her husband, who happened to be a member of the board of selectmen.

It doesn't take long for word of a scandal to reach all ears in a small community. The following day there were petitions circulating, meetings scheduled, sympathies expressed, and fingers pointed. The opinion voiced by many residents was that Elaine's action was the responsibility of the entire committee. "No committee should ever allow for only one signature on checks. What were you thinking?" I was thinking that I had never been on a committee before and would never be on one in the future. I was thinking that I should not be held responsible for someone else's actions. I was thinking how wonderful the undemocratic system is of a fishing boat. The captain makes all decisions and the crew carries them out. There are no committees, boards, or votes. Things get done.

Another point of interest, which may also be a small-town syndrome, was the revealing of the results of what I call "score-keeping and accounting." It was amazing how many people expressed the fact that they were not surprised that Elaine and her husband might have been tempted. "After all, look at the money they've spent lately. They just bought a new truck that you know they can't afford. They paid big money for that dumb dog they keep staked in their yard. And electric guitars are expensive! Oh sure, they both work, but they spend money like drunken sailors. It was just a matter of time. . . ."

The borrowed money was replaced, but old lines had been redrawn, dividing the community. The question was whether or not the town should press charges, and of course my mother and I debated it to death. The town did not pursue legal action. I suspect Ted Hoskins may have had a lot to do with the calming of those who wanted blood. Elaine resigned as chairman and treasurer of our beloved committee and received the punishment of several hours of community service. If it had been me, I would have preferred going to jail to scraping and painting the outside of the store. It reminded me of *The Scarlet Letter*.

Dave Hiltz became our new leader and we adopted a policy that required two signatures on every check.

The chatter about Elaine's mistake soon subsided. But the damage had been done. The Lighthouse Committee, once again, was front and center in the town's communal mind. Basically, we could no longer do anything right. Meetings, votes, letters, petitions . . . I dreaded getting up in the morning, knowing that I would surely encounter someone with a comment, question, or better idea. I felt I was constantly defending the committee's decisions and actions. There was, in my opinion, entirely too much unsolicited public input.

The battlegrounds were now such issues as: Who would have access to the tower, if anyone? Who would keep the key? An easement had to be worked out between the Burkes and the town, as the Burkes owned the land surrounding the tower. Easements and amendments went around and around so many times I lost count. Someone threw a wrench in the project's progress by suggesting that a nonprofit organization be formed to take possession of the tower. This triggered another slew of special meetings. "Didn't we vote on this last time?"

It seemed an eternity of uphill battles, including the one I fought daily with my mother. The town's residents finally took title to the tower, and we are all now proud owners and stewards of our most important historical landmark. More than enough money was raised for the restoration, which was very successfully completed. I've never understood why it had to be so difficult. My involvement was certainly less than enjoyable. I vowed, never again. Some of us are just not cut out for civic duty.

CHANGING THE WATER

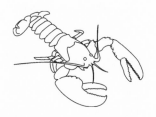

Patience is not my strongest virtue, and waiting for lobsters to make their annual appearance was trying what little patience I had. With great anticipation, Dad and I had set five hundred traps, and every morning that we chose to haul, Dad and I managed to muster both enthusiasm and optimism — this would be the day the lobsters started for the season. Today was no different.

"I'm one thousand pounds behind what I had last year at this time," Payson Barter shared quietly and matter-of-factly as he climbed into a small wooden dinghy at the town dock. I hadn't dared to figure out exactly how far behind my total was this season, but was well aware that the lobsters had missed a fairly urgent deadline set by my checkbook. Whether the lobsters were late or not coming at all was a question no one cared to pose but certainly weighed heavily on all minds. Payson slid oars into oarlocks and began to pull himself and the dinghy toward the *Perseverance,* his lobster boat. "I guess when they come, there'll be plenty of 'em . . . probably happen overnight."

"I hope it was last night," my father muttered as we made our way out to the *Mattie Belle* in the skiff. We had, in the last two days, caught a handful of shedders, and the long-awaited arrival of the new-shelled lobsters was the source of hope that kept us going, a promise that more would soon follow.

But all that followed the few trappable lobsters over the next few days turning to weeks were strings of gear brought into our area by outsiders. Jack MacDonald referred to these men as "thru-fishermen," meaning they started with the first shedders up in the Bay, and shifted their gear to other, outer grounds as the shedders came out. Like nomads or gypsies, the thru-fishermen staked no season-long claims on any particular spot, and showed no loyalty to, or association with, any group of lobstermen bound together by heritage, tradition, or history. Fishing tactics had changed, and the modern fishermen followed the lobsters. Islanders had not changed, and could not. Even the youngest Island fishermen sat and waited for the lobsters to come to the same grounds Islanders have fished for generations. Although the practice of thru-fishing is totally legal, Islanders could never be thru-fishermen, as we are restricted by our respect for time-honored boundaries. The yearly increase in the number of thru-fishermen arriving with the lobsters for which we had been waiting patiently was what had driven the Association's plan to create our own fishing zone in the first place. An exclusive zone would keep the interlopers off our traditional bottom legally. And allow us to continue to fish the way our ancestors had.

The calendar was nearing the end of July. In previous seasons, early July had marked the beginning of the lobster bonanza, 2½ to 3 pounds of lobster per trap hauled. We had, in Julys past, been assured of landing between 500 and 600 pounds every time we left the mooring. But something was different this year, and that difference manifested itself in a very poor start to what we all hoped and prayed would eventually blossom into a decent season. Hauling empty traps is often sarcastically referred to as "changing the water in the traps," and Dad and I were doing our share.

Day after day of hauling empty lobster traps is discouraging at best. We were heartened by the hopeful spirit of some lobstermen but were brought down again by a melancholy voice on the radio, despondently reporting that he just couldn't take it anymore, and was returning to the dock to try again in a day or two. It was times like these that made us so aware that Archie Hutchinson was not on the water this year. For if Archie had been well enough to swing a leg over the rail of the *Mary Elizabeth*, he certainly would have answered the dejected fisherman, encouraging him not to give up with, "Put your head down, and drive 'er!" These words were Archie's signature and he lived by them. I knew he must have been terribly sick when I hadn't seen his hot-pink buoys in the spring.

Although Archie was a Stonington and not an Island fisherman, he had never been a thru-fisherman. Islanders had shared both good and bad with Archie and a few of the other mainlanders who were more than welcomed by us, which made our exclusive zone plan even more complicated. How could a newcomer like me think I had more right to fish here than men who had been doing so since before I was born? "Birthright" Jack MacDonald had said when I posed this question to him. My roots were here, and I now lived on the Island. As poor as the fishing had been so far this season, I would have gladly given my lobstering birthrights to anyone foolish enough to take them.

Lately, I had been more than willing to head to the mooring, but was shamed by my father into staying out and continuing. "Let's use this bait up," or "I'm game for a few more, if you are," were my dad's words to encourage me to do what I knew needed to be done. The more we worked, the more I hated it. I was beginning to suspect that lobstering was not my thing. The bait stunk. The boat was muddy. And I was going broke while waiting for buglike creatures to crawl out from under rocks. What had my life become? I looked forward to the high point of my day: lunch.

The path of my existence became the flight of a bumblebee. Around and around and around we went, in clockwise circles, hesitating at every blaze-orange bloom and hovering over it just long enough to pollinate a trap with fresh herring. Endless loops of mindless, monotonous work; the rest of the world — sky, water, trees, boats — whirling by in the opposite direction, giving me the feeling of swimming perpetually against the tide. Again and again and again, I reach over the starboard side with my gaff and hook, an orange buoy. I grab the pot warp beneath the buoy with my left hand while my right releases the gaff, laying it along the rail, or gunwale (pronounced *gunell,* accent on the first syllable). Next, I pull line from the water by hand to get enough slack to run the line through the hauling block, under the fair lead, and between the plates of the hydraulic hauler mounted on the bulkhead forward of where I stand.

With my left hand, I then twist toward me the brass handle, located on the dashboard, that controls the pump that supplies hydraulic oil to the hauler's motor. The plates spin counterclockwise. The line, pinched between back-to-back plates, is pulled from the water and coils itself onto the deck below the hauler as it spins. I then toss the buoy onto the gunwale next to the house. The line strains with the weight of traps being pulled from the ocean floor, and squeaks and pops in the hauler. When the knot joining pot warp and float rope goes over the block, I slow the hauler down and watch as the first trap breaks the surface and is raised up beside me. Again and again and again the traps come up empty.

Each time, my father reaches across me, grabs the trap, and pulls it up just enough for me to clear the gangion from the block (pronounced *ganjun,* accent on the first syllable). I refer to the action of pulling a trap up and inboard, clearing line from block, and laying the trap onto the gunwale as "breaking the trap over the rail." When I fished offshore with a larger boat, we worked forty-pot trawls for lobster, and up to one-hundred-pot trawls for crabs (up to one hundred traps on a single line). One member of the five-man crew did nothing but break traps all day. That man always complained of a sore back. Now I knew why.

Dad slides the main trap down the rail as I continue to haul line, stopping so the trailer rests hanging from the block by its becket. I break the trailer over the rail and let it come to rest just ahead of the main trap. Now all the line is on the deck in a loose pile or semi-coil. Dad will take care of the main trap while I tend the trailer. Again and again and again.

As I hauled traps in Robinson Cove that day, I set them back as shoal as I dared.

Looking over the side, I watched for rocks on the bottom and had to reverse the engine to back out of one crevice I'd explored. Looking up, I scanned the land that was once inhabited by my family. A clearing where the Robinson homestead once stood was now vacant except for the old well curb. The handle of the pump was in its horizontal position, as if someone had been interrupted mid-stroke, and pointed to the tote road that lead to another small field where there was no longer any sign of the family's barn. The cleavage of the road itself was just barely discernible in the overgrowth, and I imagined it hadn't been used since we buried my grandmother for the second time.

Gram died in a nursing home on the mainland. There was no question but that she would be interred next to her parents on this hill overlooking the cove where she was born and raised. Since there is no funeral parlor, undertaker, or hearse on the Island, putting a loved one to rest has always been a family obligation, in some cases a family outing. I remembered staring at the vault (which was laid across the stern of Dad's boat) all the way from Stonington to the Island. In the vault lay a casket, in which lay

Gram, who had been waiting quite patiently for the ground to thaw, two or three months, I think.

The vault went from the boat to the back of a pickup truck, where Gram stretched out among several living grandchildren for the bumpy ride to the cemetery. I am sure this was the only occasion I got to ride in the back of a truck with my grandmother. When we reached the gravesite, we were greeted by uncles and older cousins who leaned on shovels around the freshly dug hole. An awkward and unceremonious pallbearing moved the vault into the ground, and we all took turns with the shovels. Finally the mounds of loose dirt that had been piled around the hole, and indeed the hole itself, disappeared with my grandmother. A modest headstone was set in place, and I wondered if maybe in this case it was actually a footstone. Because the vault was not marked in any fashion, we had no way to determine which end was which. What we did decide was that head or foot, it was not important enough to open the vault and casket to check. "Well, there's a fifty percent chance she's not backwards, and I feel lucky" was the sentiment voiced by one member of my family.

"I don't care what the odds are," added

another. "She's been dead for two months! Besides, if she's in there the wrong way, it won't bother her half as much as the fact that Gramp's getting married again." It was what we were all thinking. We politely ignored the person who had said it.

"Remarried at the age of eighty-three . . . Gram will roll over in her grave" was also ignored.

My grandfather did remarry, that very summer. And Gram did more than just roll over; she came completely out of the ground. I'll never forget the look on Gramp's face, or the feeling that came over me, when I wandered up to the cemetery to check it out myself the spring following the interment. There, on top of the ground, as if it had never been buried, was the vault. There was no sign of a hole or any digging. The outside of the box was clean, and beneath it was seemingly undisturbed earth.

There were several theories circulating on the Island as to what had caused Gram to rise out of the ground, box and all: Gramp's new bride, culprits exhuming bodies hoping to find family heirlooms or with other perverted intentions, or the natural occurrence of a frost-heave. I always suspected something deeper, perhaps the selling-off of inherited land. Whatever the cause, the vault was end-for-ended when it went down the second time, and has remained at rest for many years.

I needed to visit the graves, I thought, as I hauled another pair of traps from the cove. I hadn't been since Memorial Day two years before. That had been the first and last time my mother requested that Dad and I take care of planting things at the gravesites. As I recall, we had mistaken basil for something else, leaving my mother with terra-cotta pots of the something else with which to season dinner. Islanders still take care of plantings every Memorial Day, but all of the activity leading up to any particular memorial had been moved off-Island or hired out to one of the many caretakers. Hell, people didn't even die on-Island anymore. As soon as anyone reaches an age where the next birthday becomes questionable, he or she is shipped off to be cared for elsewhere. I suppose it's more convenient for the survivors to have the deceased taken care of by professionals. I wondered how long it had been since an Islander had been allowed to die at home. How long had it been since we had buried one of our own? I pondered this as I dizzied myself and my dad in starboard circles.

It is so easy to let your mind drift while

"changing the water." A pair of traps from the cove, identical to each preceding pair we had rebaited and set earlier that day, now lay on the washboard in front of me. I opened the door of the trailer and unwound the bait bag's drawstring from where it had been secured to the plastic cleat designed for the purpose of hanging bait in a trap. Opening the mouth of the bag and tipping it upside down, I shook remains of old bait into the water, where circling gulls swooped in and dove on the discards like kids on candy under a punctured piñata. Placing the empty bait bag in a tote with other empties, I grabbed a bag of fresh herring from a bucket Dad kept full. Securing the drawstring of the bag onto the cleat so that the bait hung midway between the two heads, or entrances, to the kitchen section of the trap, I eyed the lobsters that snapped their tails open and shut in the parlor. They looked small, but I would measure them just in case.

As I measured the lobsters, my father jammed herring into the two bait bags that we had emptied. Sticking one end of the brass "lobster measure" into the socket just below one of the lobster's eyes, the other end of the measure came down onto the lobster's tail below the end of the carapace: too

small. I tossed the "short" lobster overboard and continued measuring the others,
returning all but one to the sea. "Hey! A
keeper!" I exclaimed gleefully. "It's wicked
soft, too. Maybe they're finally on their
way."

Any lobster that is 3¼ inches or more in
length from the bottom of the eye socket to
the bottom edge of the shell covering the
body is legal or a "keeper." Lobsters that do
not make the measure are "shorts" or "snap-
pers" and must be thrown back. Fishermen
found to be in possession of short lobsters
are subject to fines or loss of their license to
fish. You also aren't allowed to keep "oversized" lobsters, or those whose carapace
length is greater than 5 inches. Nor can you
keep egg-bearing lobsters or "V"-notched
lobsters. A "notched" lobster is a female who
was at one time or is still bearing eggs. Fishermen cut small "V" shapes from the second
tail fin from the right (looking at the lobster's
back, head up) of egg-bearing females,
marking them as members of the brood stock
and protecting them from capture. All
notchers, even after they have released their
eggs, are illegal to sell and must be thrown
back. "V" notching has proven to be an effective measure of lobster conservation in
Maine, the only state to have such a law.

Now two freshly baited traps with doors secured closed sat on the *Mattie Belle*'s starboard rail waiting to be set back out. There were plenty of buoys in the area now, and I had to pay attention to keep them out of my wheel and avoid setting my traps over someone else's. I found a clear spot, nodded to my father, and he pushed the traps overboard one after the other, as I jogged the boat around to starboard. As the line paidout over the starboard rail, my father banded the claws of the single keeper and dropped it into a barrel that sat on the deck against the port side. The lone banded lobster splashed into the salt water and slowly settled to the bottom of the barrel. I jogged the *Mattie Belle* around Robinson Point, and headed south, down the Island's western shore.

(Banding lobsters is the act of placing rubber bands around each of their two claws, preventing them from injuring one another, or anyone who has to handle them before they're cooked. Bands are placed around claws with a "banding tool," which is shaped like a pair of pliers. A relaxed band is placed over the two tongs of the non-handle end of the tool. When the two legs of the handle are squeezed together, the tong end opens up, expanding the band to whatever size is needed to stretch over the claw being banded. Once the band is around the claw, the handle of the tool is released and the band contracts tightly around the claw, forcing it to remain closed. A lobster's claw strength is in closing, not opening. Bands made for hard-shelled lobsters have a tendency to tear soft shells; thus different bands are made for the different shells. Bands also come in a variety of colors. I do not know why. Years ago, fishermen "plugged" lobsters rather than banding them. Small wooden pegs were driven into the thumb joints of each claw, jamming them closed.)

For the rest of the morning, we hauled and hauled, around and around to starboard so many turns that I felt it might become necessary to turn to port in order to unwind. Trap after friggin' trap was hauled, picked, baited, and set, and on occasion, a keeper would splash into the barrel. My back ached, and I was hungry. It seemed to take lunchtime forever to approach, but when the clock on the GPS finally read "12:00," I was not shy about putting the gearshift in neutral and declaring it time for a sandwich.

I stepped over to the port rail to clean some of the mud from my hands and oil pants with the deck hose. Somehow I summoned the courage to look into the barrel. I could see pieces of its blue plastic bottom down through the lobsters. "Shit," I whispered, not quite under my breath. To make matters worse, I had packed our lunch this morning. My mother normally made lunch for us, but she had slept in today. It's embarrassing for me, at the age of forty, to admit that my mother still prepares lunch for me every day, so I usually don't mention it. Sleeping in . . . my mother was indeed showing signs of her age. She seemed to require more sleep than she used to — or maybe it was making lunch that was getting old. Mom had always had subtle ways of putting things into perspective. I silently vowed to tell my mother how much I appreciated her. Maybe she would make lunch tomorrow.

As my father washed bait and slime from his gloved hands, I dug into a small Igloo cooler that served as our dinner pail. "Can of sardines, Dad?" I offered.

"No. Do you have a sandwich in there?"

"Well, not exactly."

"What exactly do you have?"

"Sardines and peanut butter crackers." My father joined me in eating what we both considered a damn poor excuse for lunch.

136

We sat on the gunwale, side by side, his feet on the deck, mine dangling 6 inches above it. We ate prepackaged crackers one shade off my buoy color, with a dab of peanut butter between them, and washed them down with warm soda. The sardines were unappealing even to me, and the unopened cans remained in the cooler. We sat and ate silently.

As we ate and drifted, I watched an island kid named Jason hauling traps by hand from a skiff. Hand over hand, he pulled until the trap broke the surface, then yanked it onto the gunwale. He would, I thought, graduate to a larger boat by the time he finished high school, as his father and grandfather both had done. It was young people like Jason who would keep the Island alive. He would marry and have children. His children would follow in his footsteps. Now Jason hauled in the sun's blaze, a stick figure in a black skiff, moving purposefully along the shore. The scene was the perfect accompaniment to a peaceful lunch. Peaceful, until the roar of a powerful diesel engine and a puff of black smoke came from behind the small peninsula of Trial Point at the mouth of Moore's Harbor. Dad and I stared and waited for the source of sound and exhaust to round the point. High white bow, full midships, and finally a wide stern appeared from behind the spit of ledge. This boat, much longer and beamier than any owned by Islanders, carried on board a mountain of new lobster traps that melted out of sight as the two men in the stern hustled to throw them over the side, setting them in close to the shore.

"Another factory ship," my father remarked of the large boat. A boat with two sternmen was very productive, with the ability to haul many more traps in a day's time than a boat with one sternman. That was another advantage to fishing from the mainland, I thought: workforce. Many Islanders fished alone simply because there was no one looking for work. Unless a father had a son without his own boat, or in my case, a daughter had a retired father willing to work, the options were few. Fishing alone seriously limits the number of traps that can be handled from a boat, which limits the number of pounds that can be landed. Occasionally, a stranger would appear, seeking employment in the stern of a boat. Strangers never stayed long, as the disadvantages of Island life overwhelmed the advantages, and the work proved harder, dirtier, and less romantic than they ever imagined. I had my own experience with a stranger seeking employment the year before, and learned not to expect much.

The sharp contrast of Jason balancing in a skiff against the hull of the much larger mainland-based vessel was a nudge of sorts. Suddenly lunch was less than idyllic. When its pile of traps had all been set, the big boat steamed off without a glance our way from any of the three men aboard, nor did the captain take care not to capsize the small skiff that flopped rail-to-rail in its wake. Jason barely held on.

STERN-FABIO

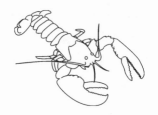

As I sat watching Jason resume hauling, after the "factory boat" sped off, I remembered with mixed emotions my short career as a "factory boat" captain, or rather as captain of a lobster boat with two sternmen. It had been just one year ago that I had tried this, and it began on a gorgeous summer day when the lobsters were crawling into traps in great numbers.

I had arrived at the town dock at the same time as the day's last mailboat run. There were swarms of people coming and going. Well, maybe "swarms" is a bit of an exaggeration, but everything is relative. In any case, there were a lot of people around. Well, let's say twenty people. Cars and trucks jammed the dock and parking lot. In the crowd, I couldn't help noticing an extremely handsome young man leaning against the dock's railing seductively smoking a cigarette. There is really nothing sexy about smoking, but he did have a certain allure. I have always been quick to admire attractive people, but this guy was different. He stood out in all the activity by being inactive. He was cool and relaxed while those around him scurried with luggage and boxes of groceries up and down the wharf. As I passed by him on the way to my truck, he surprised me by asking, "You're Linda, aren't you?" We shook hands, he introduced himself as a friend of a friend, and explained that he would be on the Island until late fall and was in need of a job. From his hairstyle and clothes I knew he wasn't local, and wondered what he knew about working as a sternman aboard a lobster boat. He was well-spoken and polite, and I learned from our conversation that he was college-educated and on a leave of absence from a job in Portland. I told him that I was not in need of another sternman, and wished him luck. He persisted, politely, saying that he was anxious to try lobster fishing, and that he was willing to work for no pay if I should care to do him the honor of giving him the opportunity. He was begging. Now I was certain that he had never stepped foot aboard a working commercial fishing boat. He said that he would be living in a tent pitched on our mutual friend's property, and that his needs were few. I liked him and hired him with the understanding that he would be working with my father and me, an old man and somewhat fragile woman (for "fragile," read "lazy"), both of whom had sore backs. I made it clear that I would expect a strapping young guy like him to do the majority of the heavy work (for "majority," read "all"). And in return, he would be given a full 20 percent of all proceeds from lobsters caught when he was aboard. (This was and still is the going rate for a sternman, and I was more than happy to share my money with someone who could make my father's and my life easier.) The young man was clearly gay, and this would raise the Island's tally to four, giving our small population a gay/straight ratio that might rival Key West's. I include his sexual orientation not simply as a point of information but in my own defense. Otherwise, I suspect that I might be accused of hiring a sternman I didn't really need solely because he was strikingly handsome and seemingly unattached.

Within the first ten minutes of hauling traps with a third person aboard, I wondered why I hadn't hired someone before. The new guy was a quick study, very strong, and great company. He enjoyed the work and was quite vocal in thanking me for the job. Hauling traps was much faster and easier with his help. We were able to haul and set many more traps in one day's time than we could alone. And our backs were fine. Dad liked him, too, and agreed that we were making more money with the young man than we had without him. Handle more gear, catch more lobster, make more money . . . even after the 20 percent paid to the helper. The more we got to know him, the more we liked the man. Lobstering was fun.

After a week of successful fishing and perfect weather, the man requested two days off; another job opportunity had presented itself. He had been asked to work as a model for an art class in Stonington. The pay was fifty dollars an hour. I was happy for my new friend and agreed that we could certainly work around his schedule. Fifty dollars an hour is unheard of in our area, and I wondered why some Stonington fisherman had not jumped at the chance. When I learned that he was to pose nude in the granite quarry, I understood, and that's when he became "Stern-Fabio." I prayed for warm weather and slow students, on his behalf.

As soon as the modeling began, Stern-Fab became less than reliable. It was obvious that as long as he had money in his pocket, he had neither desire nor ambition to work. I guess if I could make fifty bucks an hour lying around naked, I, too, would prefer it to handling bait, mud, and slime. (Actually, a friend did offer me the same money for the same work, but only wanted to hire me for ten minutes.)

A few days later, after promising to meet us, Stern-Fabio stood us up. We waited that morning like children for a deadbeat dad, hoping that he might just be running late. We gave him the benefit of the doubt, but finally had to leave without him for what turned out to be our best haul of the season. I felt bad that Stern-Fab had missed such a nice paycheck. That night he phoned and apologized profusely, saying he had overslept, and vowed that it would never happen again.

But it did happen again, the very next day. Again Dad and I waited at the dock, hoping he would appear. He did not, so we fished without him. He called that night, wondering why we had left early, and claimed to have arrived at the dock at the designated time, only to find the *Mattie Belle* long gone.

I was relatively sure that my father and I could both tell time, and even more certain that both of our watches could not possibly be wrong. I asked Stern-Fab to synchronize his watch with "our time" and told him to be at the dock at six A.M. the following morning, explaining "that's when the big hand is on the twelve, and the little hand is on the six." I normally would not have been as tolerant, but I did genuinely like the man.

As the next few weeks unfolded, Stern-Fabio came up with all the same excuses and apologies that I had heard in the past from errant crewmen. It was disappointing that he could not even think of an original excuse for being a no-show. He suffered everything from a dead grandmother to a broken car. Dad and I pretended that we simply didn't care. But of course we did. Whenever Stern-Fab decided to grace the deck of the *Mattie Belle*, we gladly took him fishing. Most of the time he didn't, and we had to leave the dock after some hesitation.

One day Stern-Fab made a surprise visit to find me up at my folks' place. He had a trip planned and was seeking a paycheck with which to fund it. I never had any shortage of work to do, and liked the fact that someone wanted to help. It is difficult to hire Islanders because they have more than enough work to do on their own. Rather than take him out fishing that day, I put Stern-Fab to work cleaning the trap lot. Now that the traps were all in the water, the area looked a bit deserted and in need of a sprucing-up. When the job was done, and the time came for Stern-Fab to sprint for the last mailboat, I paid him cash. He hurried happily off to Bar Harbor, where he was excited about taking in a drag show. I guessed that although he promised to return to go fishing the next morning, I would not see Stern-Fab for at least three days and wished him a good time.

The following morning I received a call from an ex-friend who wanted to know why I had called and hung up on him three times the night before. I was confused. This voice on the phone that I had not heard in weeks was not speaking to me the last I knew because I had inadvertently killed all the fish in his aquarium. (It was an accident and I was *very* sorry. He didn't actually hate me until I referred to his saltwater exotics as "guppies.") I hadn't phoned the fish mourner, and told him so. He reminded me that his cell phone had been in my truck for some time, and that it was programmed to ring his home number whenever anyone pressed "send." His home phone had caller

ID, on which the cell phone number had appeared too late at night for decent people.

I explained that the truck with his phone was parked in Stonington at the mailboat dock, and that I had not been to the mainland in a month. Some drunk must have been sleeping it off in my truck; I promised to take care of the problem and return the phone. I immediately called my friend Clarence Oliver, who works at the mailboat dock and is kind enough to do me an occasional favor. I asked Clarence to check my truck for drunks and damage. "Your truck ain't here. Some blond guy with lots of muscles took it yesterday."

Stern-Fabio had stolen my truck. He was somewhere between Stonington and Bar Harbor, doing who-knows-what in my vehicle. And he was using the phone. No doubt trying to call me with an excuse to not go fishing. Now I was mad. I put out an all-points bulletin (Clarence and the mailboat captain) that as soon as the thief surrendered my truck, I wanted to know. I couldn't wait to fire him, and would meet every mailboat until he surfaced. I planned to really come unglued, and that plan had caused quite a crowd (lynch mob) to meet the boats with me.

We were all disappointed. Stern-Fabio

did not appear. Several people suggested that I "call the cops," but I didn't want Stern-Fab arrested. Eventually I got word that my truck had been returned to the dock safe and sound, and that the handsome blond guy did not stick around long after Clarence informed him of how I felt about my truck being borrowed. The jig was up. I suppose he thought he'd get away with borrowing my truck without my permission. Funny thing is, he would have if he hadn't been so stupid as to use the phone.

GEAR WAR

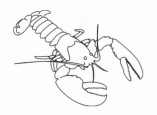

I knew the time had come for me to attend a meeting of the Island Lobstermen's Association when I watched a mainland fisherman set a string of gear north of the Sawyer Buoy. The one tiny piece of sacred lobster ground not fished by outsiders is bound on its southern end by an imaginary line drawn from the green navigational can-buoy between Robinson Point and Kimball Island (Sawyer Buoy) to the lighthouse. This water is fished exclusively by Islanders. A relatively puny piece of water, it's all we have managed to hold on to, and now it had been invaded.

Five foreign buoys marked the ten traps, or one string, that had been set not in the gray area near the imaginary boundary line but brazenly smack-dab in the heart of our sacred puddle. The presence of the unwelcome gear was insulting and showed a total lack of respect for tradition. The placement of the buoys was the throwing down of a gauntlet, a challenge to go to war. I, for one, was ready.

The same string of gear, fished by the same instigator, had been politely moved back over the line for him just days before. He had been persistent, and set the string even deeper into our territory, right in our dooryards, as it were: a flagrant violation of unwritten law. He had been warned. Now was the time to "cut him out of the water." This was the method widely used, accepted, and understood. It is the method that keeps us from fishing Head Harbor and the entire south end of our own Island. (That's right. A family of mainlanders have maintained a large area for themselves, keeping all others out with a sharp knife. We are not allowed to fish the water we can spit into from our southern shores, because it has been and continues to be fought for and protected by a dozen men who commute quite a distance to do so. It's just the way the business works. So, to my thinking, it should have worked that way for us, too.) As luck would have it, this was the third Wednesday of July, the evening members of the Island Lobstermen's Association reserved for the monthly meeting. I was indeed ready.

After two years of meeting, planning, scheming, voting, and working toward legislating an exclusive zone for us to fish, the members had backed out of the effort gracefully enough last winter by claiming to be satisfied with the status quo. Actually, it was the wives who eventually nixed the plan. I understand from my father that the special meeting that included any interested resident brought women who were vocal about what they thought would happen should we succeed in legally booting mainlanders from the most coveted territory around. I was off-Island at the time, but heard that the women feared retaliation by those excluded. So two years of work was dropped like a turd. After I recovered from both shock and disgust, I remembered that legislation had not been my first choice anyway. I had preferred waging a gear war, and tonight's meeting would be the perfect time to muster the troops.

I fidgeted through the first hour of the meeting. Minutes of the last monthly gathering were read, the treasurer reported all current figures, and a number of other boring things were discussed. Everything from the scarcity of lobsters to the price of bait was gone over quite thoroughly while I anticipated the beginning of our war tribunal. Should we buy a new valve for the hydraulic winch? The question was debated to death. Do we need to buy more shedder bands? Which ones do you like better, red or yellow? Hell, if we didn't hurry and go to war, we wouldn't need any bands at all, I thought. Finally, just when I thought I was once again wasting my time at another meeting, Jack MacDonald asked, "What do you want to do about the gear above the Sawyer Buoy?" That's all it took. The place lit up like a tinderbox, and Jack had struck the match. Everyone chimed in with some sentiment on the importance of sending a message through the treatment of the intruder's gear, and the talk escalated perfectly to how we should be "trimming up" some others, too.

Jack crossed his lanky legs, rested an elbow on his top knee, and dangled his wrist so that his cigarette was in the proper position for a drag. He watched and listened. Jack's legs were long enough to take two full wraps around each other, one at the knee and one at the calf, and still have enough slack to rest both feet flat on the floor. Other than the length of his legs and the ever-present cigarette, particulars about Jack that could not go unnoticed were his eyes and hands.

If Jack had not despised politicians and lawyers, he would have been successful pursuing either career. His blue stare cut through the curtain of smoke that rose from what was left of the cigarette he always held loosely between yellowed fore and middle fingers. Jack's gaze was so painfully intense when he spoke, and I was so taken with his eyes, that I never saw his mouth move. I was relieved when he looked away from my direction, releasing me from some strange optic constraint. He had listened, and it was now his turn to speak.

Jack's fingers, like roots, appeared to have the ability to grip and hold fast even the most tenuous strand of hope. I believe that his hands were the symbol of his entire being. When Jack latched on to something, he wasn't apt to let go. Islanders first learned of Jack's tenacity in the 1970s. At that time, the National Park Service had set its sights on the Island with the intention of obtaining the entire rock for a national playground, relocating families of Island dwellers to the mainland, where the displaced fishermen could find work pumping gas or selling shoes. Jack had a different plan for the Island; he led a passionate fight against the federal government and won. If anyone could lead us into war against the mainland fishermen, Jack could. The challenge for Jack would be to mobilize this ultra-conservative, moderate-tempered, peace-loving membership.

Jack said all he could about the hard facts that should have riled us to the point of storming out the door and onto our boats to rid the entire ocean of all traps not fished by us this very night. He eloquently laid the groundwork by pointing out what we already knew. Through the years, Islanders have struggled to keep their lives simple, with fishing as the economy base, rather than making tourism the Island's mainstay. But with the Island depopulating quickly, the key to our future lay in protecting our current grounds and securing new ones. The biggest threat to our way of life wasn't tourists — it was too many traps competing for the resource that was needed to sustain our fishing. Traps were going to have to be removed from the water. And they weren't going to be ours. Jack smoothly bent the debate toward the idea of a gear war. I couldn't have imagined a more perfect approach.

Wars begun with traps cut out of the water had, in the past, escalated to violence in the form of burnings, sinkings, even shootings. How exciting, I thought. As the discussion heated up, I imagined my part in the battle. I would gladly sacrifice all my gear, my fishing license, even the *Mattie Belle* for the cause of saving what was left of our precious Island life. My father has a permit to carry a gun, and I knew he would not hesitate to use it in self-defense should it become necessary. I would sleep aboard my boat every night, and would hear the engine of anyone trying to sneak in under the cover of darkness. We would post remote areas with sentries armed with radios. We had five certified scuba divers in the Association. They could swim along the bottom and sabotage outsiders' gear and not be spotted by the marine patrol and wardens who would be forced to try to stop the war. The Coast Guard would be called, but we had the advantage of living here! How could they catch all of us? The damage would be done before anyone caught on to what we were up to. Finally, the opportunity to live out my pirate/outlaw fantasy was coming of age. We would need to work together and quickly to cut as much gear as possible in the first attack, I thought. We would start tonight, I imagined. By tomorrow morning, the ocean would be littered with buoys drifting aimlessly, their warps severed by knives clenched in fists of enraged Islanders who had finally been pushed over the edge. We would no longer be considered door-mats. We would be feared. We would be regarded as crazy. It was perfect. The activity of my overzealous imagination slowed as I tuned back into the meeting.

"Fishing's dangerous enough without worrying about being shot out of your boat."

"Cutting gear is illegal. What if I get caught?"

"I can't afford to lose any gear."

"I can't afford to lose my license."

"I could never touch another man's gear. Just couldn't do it."

"My wife would kill me."

"There's plenty of room for everyone."

"The lobsters will be here soon."

"I'm late for dinner. I make a motion to adjourn."

"Second."

Dumbfounded, I sat and watched the group funnel through the door and spread to various vehicles. Jack sat, appearing as cool as the underside of a pillow. Reaching down, he ground the end of his cigarette into the floor.

THE FIRST CASUALTY

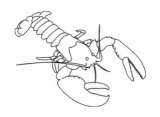

Waking before dawn, I was disoriented. I was unsure of where I was and my heart pounded rapidly. Around me was silence. Perhaps the engines had died while I slept, and the boat had been drifting aimlessly. What if the batteries had been drained so dry that they would not turn the starters over? How long had I been asleep? Where was I? Slowly my eyes adjusted to the lack of light, and gray blobs took shape and came into focus. I was relieved to know that I was not aboard a boat and would not spend the day kneeling in the engine room bilge, turning wrenches and tearing knuckles in shadows cast by a flashlight held by a nervous crewmember. Climbing out from under the quilt, I sat on the edge of the bed in my parents' house and willed my blood pressure back down to normal.

I wondered what exactly had scared me. I concentrated, but failed to bring back whatever I had been dreaming about. Fear, in this case, had something in common with a second cousin twice removed; there was a connection there somewhere, but it was twisted and convoluted. My nightshirt was stuck to my back. I had been sweating. What a nightmare.

I have never understood why some things scare me while others don't. I had spent all of my adult life engaged in the world's most dangerous profession, commercial fishing, but never feared the inherent, well-known, physical danger. My gravest concern when leaving the dock had always been that I would catch nothing, not that I might never return. When I first made the decision to go offshore, I was more frightened about what my mother's reaction would be to my career choice than I had ever been of perishing at sea. And as I vividly recalled that twenty-year-old scenario, my mother had not disappointed me.

"Fishing? What about law school?"

"Oh, I'm putting off law school for a year," I lied in an attempt to deflect the anger I could see mounting. She drew a deep breath through flared nostrils, and held it for what seemed an eternity. Her face grew red under raised eyebrows and around a tight-lipped grimace. She hesitated, thinking, planning her attack. I braced myself as I watched her eyelids snap wildly open and shut, and wondered if she would ever exhale. Which, of course, she did. As if she had been struggling toward daylight at the surface of the water above her, after being held under too long, she exhaled violently. Now her breath came in short, erratic gasps. My mother, who had a passion for drama, would soon hyperventilate.

So I wasn't that surprised when my mother flung open a cupboard door and grabbed a stack of dinner plates. It is what she does when she is very upset. *Crash!* I had seen her empty the contents of every cupboard in the house, hurling plates, glasses, cups, and bowls to the floor where the shards of pottery and glass formed a kaleidoscope at her tiny feet. Once everything within reach had been shattered, she launched a verbal assault that I believed to be one of her best. There was no sense trying to interject anything in self-defense while she delivered the barrage, nearly hysterical. The screaming continued as I made my way to my bedroom to collect the duffel bag packed with everything I would need for thirty days at sea. The high-pitched voice was right at my heels as I found the front door. Then, realizing that I was indeed about to leave, the tone switched to more of a pleading. My mother said all that I expected she would about responsibility, obligation, and throwing away a bright future, none of which fazed me. Slinging my bag into the trunk and climbing behind the wheel of my car, I was seconds from a clean getaway when she fired her most damaging shot. "Your father will be so disappointed."

In a close-knit family such as mine, disappointing one's parents is a far greater sin than simply infuriating them. At the age of twenty-two, I had given my folks plenty to be proud of. "Overachiever" was the term most commonly used, and perhaps my desire to succeed was driven, at least in part, by a need to please my family. Excelling in both athletics and academics, I probably gave my parents hope that I would continue along a path that they deemed worthy. Although it was important to me to keep my parents content and the dishes intact, I have always been selfish in putting my own happiness first. Scenes of flying coffee mugs cause a child to question her parents' sincerity in the repeated proclamations of "all we want is for you to be happy." My mother had already lost credibility when she played her final card and cried.

Twenty years later, my mother still does not fully appreciate my chosen career. I suspect that introducing me to people as "my daughter the fisherman" was easier now than it had been at first, but she has never gotten over the worry and fears that go along with knowing a loved one is at sea. Fishermen are not generally fearful of what others would assume to be paralyzingly scary. Islanders, too, I thought, as I made my way downstairs, keep a strange distance between themselves and apprehension of real danger. We fear more the collateral consequences than danger itself. We do not fear burning to death, but we do fear that the absence of a legal fire department may prohibit affordable homeowners' insurance. We do not fear sickness or injury but we do worry about troubling our neighbors when we need evacuation to a medical facility. We do not fear death by drowning — in fact, many Islanders never learn to swim — but everyone loves to speculate on how long a body can survive submerged in certain water temperatures. (If you can't swim, what's the difference?) Some innate

Islandness keeps our fears of physical danger from impinging on our lives. So I was beginning to think that the reason most of the Island fishermen were hesitant to fight for the rights to our own fishing grounds was not fear of physical retaliation but the sheer inconvenience of an all-out gear war.

As I loaded the coffeepot, I still had that strange feeling, the aftermath of having been scared. There was comfort in being surrounded by familiar things, and seeing everything exactly as it had been the night before, prior to sleep and bad dreams. The ever-present cribbage board and playing cards rested on the table where Mom and I had left them after counting our final hands last night. Cribbage boards are as common as wooden spoons in the kitchens of this island. Children raised in this part of Maine often learn to play cribbage before they can tie their own shoes. Some people might find that backward, but it's actually quite advanced. Long before I knew that one plus one equals two, I understood that seven and eight are equal to fifteen, and that fifteen is good for two holes on the cribbage board. Now you know why pull-on rubber boots are so popular in small-town New England. Stepping into mine, I headed to the post office, knowing that by the time I returned, the coffee would be ready and my folks would be up.

The section of rutted dirt road, just before the pavement starts, was always the blackest and creepiest part of my teenaged treks home after a dance or late-night card party. These few hundred feet had been capable of conjuring up every ghost and spirit that otherwise lived only in the stories told to convince children to be home before dark. I remembered telling my youthful self there was nothing to be afraid of, and vowing not to break into a dead run with the first rustle in the trees. I do not recall a single trip when my heart wasn't suddenly in my throat. A startled rabbit skittering from ditch to ditch like a bead of water in a hot skillet would pace my heart at the same beat. Inevitably, I would run, and upon reaching safety would be exhilarated and think myself foolish for being scared in the first place. Fear had never been a totally unwelcome emotion, and it seldom found me home before dark. I have now experienced fear many times in adulthood, mostly at the hands of severe weather at sea, and know that it is the feeling of relief, not the danger itself, that is seductive. I have always found the aftermath of having been scared a desirable state.

This morning I tried to conjure up in my mind the image of Walter Rich, the main character from a childhood ghost story that always succeeded in scaring me shitless. The ghost of Walter Rich had allegedly been tramping the Island since his drowned body was found washed up on the beach fifty years ago. I used to imagine him as a dark, faceless figure, draped and dripping in seaweed. I reached the main road without producing so much as a single goose bump.

I walked by a small clearing that held a blueberry patch the kids had not yet picked bare. Small and delicate, these berries are astonishingly resilient. The ones in the ditch thrive on dust from the road. "Take those home and wash them before eating," some mother might say, halting a small hand between Dixie cup and mouth. This morning's dew clung to the berries like syrup, moistening their skins to the color of mussel shells still damp and freshly exposed by an ebbing tide. Blueberries and recently triggered thoughts of my childhood freshened my new and growing fear that I would never have any of my own. (Children, not berries.) The fear of living alone, dying alone, and leaving no blood legacy looked increasingly likely. Who would inherit my birthright to the Island's lobsters and blueberries? Then it came to me quite clearly. "Wrecked on the lee shore of age" was the last line I had read before drifting off to sleep. A line from a Sarah Orne Jewett novel had struck a nerve and ruined an otherwise peaceful night. Or maybe it was that, combined with lingering anxiety from a mystery that had preoccupied all of us over the last few days.

It had all started that Monday. My father and I were hauling traps — still just changing water, but keeping busy. The VHF radio was on — as always, background noise. We almost missed hearing that a boat just to the west of us had been found going around in circles with no one aboard. My first thought was that its owner was dead. Next, I strained to hear the name of the boat to see if it was a mainlander's or Islander's, and whether I knew the missing lobsterman.

I remembered through the years reading newspaper articles and seeing local television news spots about lobstermen drowning. Sometimes a body is recovered, but not always. Usually the victim was fishing single-handedly, nobody to help him back aboard, nor any witness to what had happened. Ben MacDonald's father had been one such victim, drowning when my friend Ben was just an infant. The man who was now being searched for could have simply fallen overboard while reaching to gaff a buoy. More than likely, he had become entangled in his own gear while setting it. Two traps sinking quickly to the bottom, with a boat steaming away from them, a coil of line jumping from the deck, a half hitch of wet rope snaking around an arm or leg could, and had, yanked men from their feet, over the rail, and into the water in a heartbeat. Once in the water, with the boat steaming away, anyone would panic. With the weight of the traps plummeting toward bottom, would a trailing man be able to untangle himself? Even if he had a knife, who would have the presence of mind to use it and cut himself free?

I have heard that hip boots and oil clothes will kill even the strongest swimmer. Then there's the temperature of the water. How long does it take for all life-sustaining systems to shut down when submerged in 50-degree water? The thought made me shiver. My greatest fear while fishing offshore was of falling overboard at night while my shipmates slept. The reoccurring scenario in my imagination always had me struggling to remain afloat in the wake of the boat. Every time I would try to scream, a cresting wave would hit me in the face, muffling any chance I had for rescue. The lights of the boat would slowly fade and disappear while I wondered how long it would be before someone discovered me missing. Ten miles to the west, a man was drowning, or had drowned, living out what had once been my worst nightmare. Within minutes of the first radio report, the circling boat was being towed to the dock, and many had joined the search hopeful to find the fisherman patiently treading water. It was not, it turned out, a boat I knew — but I felt sick with worry all the same.

Through the monotony of filling bait bags, Dad and I listened to the radio and learned that the men searching for the lost fisherman were now hauling his traps on the assumption that his body would be found snarled in his own pot warp. The speakers again fell silent, and I figured that by the end of the day we would hear the sad news that a corpse had been recovered. I wondered at that time if the man, suspected dead, had a wife. I wondered about his children, if he left any, and how old they were. I wondered if there was a special god for widows and orphans. I wondered if his family even knew what was happening, or if not, who would tell them. Was his wife preparing dinner and threatening bickering children with "Wait until your father gets home"? My biggest worry, that my fishing grounds were not as expansive as I wished, was petty compared to what the widow across the bay would have to face. I was glad we had called off — at least for the moment — our war. It's at times like this that one remembers that the ocean, when she wants to be cruel, doesn't distinguish between mainlander and Islander.

Now I heard on the VHF that this lobsterman's gear, along with everyone else's in the vicinity, was being removed from the search area to enable the authorities to drag for the corpse. Anyone wishing to remain optimistic at this juncture was confident only that a dead body would soon be recovered. It seemed a grim thing to hope for — that oft-used word, closure.

Nearly five years had passed since the Halloween Gale of 1991 had claimed the lives of six friends and fellow swordfishermen. No bodies, not even a trace of the boat, had ever surfaced. I wondered whether anyone in Gloucester still imagined the *Andrea Gail* washed up on the shore of a deserted island; her crew living like Robinson Crusoe. Family members bent on this fantasy had been accused of denial. In my opinion, ever since the conception of the first twelve-step program, denial has taken a bad rap. Denial may be the best way for some of us to cope with loss. When I close my eyes at night, I take delight in seeing my old friends from the *Andrea Gail* drinking from coconut shells.

The radio mystery continued into the next day, when boats dragging nets behind them and men donning diving gear failed to produce a corpse. As Dad and I lingered on the mooring late that afternoon, cleaning the *Mattie Belle* more carefully than usual, we were saddened to hear that the search had been given up, and I wondered how long it would be before the body of the lost man washed up on a shore somewhere, left by the tide like a piece of driftwood. Just before turning the battery switch to its "off" position, we overheard the final radio report. When the proper officials had hauled the missing man's boat from the water for inspection for a quick sale to benefit his family, the remains of a body and clothing were discovered jammed between, and wrapped around, the rudder and propeller. It was chilling to hear, even through the speakers from across the bay.

Death at sea is all a part of this life we have chosen, I sometimes think. It's part of being a fisherman and part of being an Islander.

As soon as I heard the first radio transmission of the tragedy unfolding across the bay, I thought back to 1983 and an accident that rocked the Island. I had not thought of it for years.

OLD WOUNDS

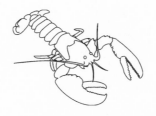

In May of 1983, five young Islanders decided to go "to America," as we jokingly call the mainland, and take in a movie. Whimsical field trips off-Island were a sure sign of spring. After "hunkering down" for long months of a winter that becomes tedious between the sparse rituals of Christmas cookies, spiked eggnog, and the annual town meeting, the thought of traveling to Stonington's opera house to see Tom Selleck in *High Road to China* must have been quite appealing. Exercising their right to just "do something," the five Islanders probably felt like they were coming out of mothballs, seeing a different light and smelling a different air for the first time in a long while was always exciting. The weatherman called for nothing more than a breeze out of the southwest, and good visibility except for the possibility of a brief passing shower. So the five set off in a boat to see the flick; only two would return.

I was at college at the time and was both devastated and obsessed with the accident. My obsession came, I suppose, from two fronts. First, two friends of mine were involved. One had died along with two of his friends, and the other had been one of two to survive a living nightmare of such proportions that no one else could ever possibly perceive or attempt to relate to it. (I knew the other two who died, but not well.) Second, the tragedy had occurred on the ocean, where I planned to spend the bulk of my time in the years to come. Soon enough, though, I became consumed not with the tragedy but by the amazing story of survival. This obsession took the form of a hunger to learn all that I could about the survival. Why had two people been able to endure such deadly conditions? A series of reported facts strung together by researched possibilities and probabilities became my very own personal script that fulfilled a need to cope, explain, and justify. The description that follows is largely speculative, and because the incident is never discussed openly on the Island, my writing of it may be viewed by some as opening old wounds, and for that I apologize.

There was nothing unusual about this trip to see a movie, except for the conclusion. What caused the skiff to capsize, resulting in five people immersed in 42-degree salt water, we'll never know. Joseph Conrad would blame Mother Nature: "The ocean has the conscienceless temper of a savage autocrat spoiled by much adulation." This May 7 of 1983, the ocean arbitrarily flexed a muscle, dumping five unsuspecting people into the frigid water off Green Island, one mile into their trip home.

I learned through research that a plunge into cold water can cause hyperventilation, which can lead to confusion, muscle spasm, and eventually loss of consciousness. This sudden immersion may no doubt also result in shock, and of course hypothermia. Hypothermia is defined as a reduction of the core body temperature below the normal 98.6 degrees Fahrenheit. There are three grades of hypothermia — mild, moderate, and severe — defined by temperature ranges to which the body's core drops. The initial symptoms of hypothermia are not specific, as cold tolerances vary widely among individuals, but there may be weakness, drowsiness, lethargy, irritability, confusion, and shivering. A formal explanation is that hypothermia occurs when the body loses heat faster than it can burn fuel to replace it. After the initial shock of contact with cold water, the onset of hypothermia and its symptoms are subtle: movement becomes slow and clumsy; there is a general lowering of physical coordination; reaction time is longer; the mind can be blurred; judgment impaired; and, in some cases, hallucinations occur.

Heat is lost from the body to the environment by conduction, convection, radiation, evaporation, and respiration. Loss of heat by conduction or direct contact is five times higher when in wet clothes, and up to twenty-five times higher when immersed in cold water. Water has a great capacity for thermal conductivity, and a person submerged in cold water can transfer heat from skin to water a hundred times faster than into air. The actual rate of core temperature drop fluctuates from victim to victim, and is affected not only by water temperature but also by clothing, initial core temperature, gender, fitness, drug use, nutritional status, and behavior in the water.

The human body generates heat through metabolic processes, which are maximized when involuntary shivering takes place.

Within varying times of exposure, because of exhaustion and depletion of muscle energy supplies, shivering stops. When the ability to shiver ceases, the cooling process becomes even more rapid. At this point, skin, surface fat, and superficial layers of muscle are the only deterrents to a worsening of the hypothermic condition, as they act as an insulating shell for the vital organs.

In the most advanced stages of hypothermia, a victim may become comatose. Breathing may stop. Pulse and blood pressure may become unobtainable, leading others to believe the victim has died. I have read many cases where victims, presumed dead, have made miraculous recoveries. One source in particular states this phenomenon quite bluntly: "No one is dead until he is warm and dead." I was relieved to learn that hypothermia dulls the sense of, and reaction to, pain. When death does occur due to hypothermic conditions, it is a sudden death from cardiac arrest or, common in immersion situations, death by drowning.

The U.S. Coast Guard has literature on hypothermia, saltwater survival tips, and treatment of victims who have been submerged or exposed readily available for mariners. Perhaps the most enlightening statistics are found in a "hypothermic chart." These charts, some of which are in graph form, basically show the expected time of survival when one is immersed in water of different temperature ranges. Although the information varies somewhat among the different sources, the findings for survival time in water temperatures in the 40- to 50-degree range are frightening. In 42-degree water, the average adult may become totally exhausted in as few as fifteen minutes, and may be expected to survive from only one to three hours. These statistics are common knowledge among most people who work on the water, and are the reason many fishermen cite for never having learned to swim. "If you're doomed when you hit the water, why bother prolonging the agony?" But there are always the few exceptions, the odd individuals who defy scientific fact and medical research by not dying within their allotted statistical time frame. And tales of these rare and stubborn individuals is where all hope lies for those who search and wait.

So, one mile into their trip home, just south of Green Island, a combination of wind and tide probably heaved up more of a sea than was anticipated. I'm guessing the outboard motor died when water came aboard over the stern. The boat may then have come side-to the seas, filled with water, and then capsized. Picture five life jackets and the skiff's gas tank immediately blown out of reach and quickly vanishing into the darkness. The five would have huddled around and clung to whatever was available to them, a small point of the skiff's bow that bobbed above the surface. Initially stunned, then quickly evaluating options, and no doubt silently weighing chances, their senses of time and distance must have become warped. I imagine there are two clocks, one at either extreme of the realm of what we consider conventional time: The painfully slow clock deliberately marks progression one wave at a time with each loud tick of the secondhand. The other clock, an hourglass of accelerated velocity, measures life that is slipping away too quickly, never makes a sound, yet can't be ignored.

The bulk of the skiff, kept submerged by the weight of its own outboard motor, would have been subject to the tidal currents, which probably teased the survivors swept along with the helpless boat as it carried them between and by several small islands that littered their northeastern drift, but refused to make landfall. The relative quality of distance must have become exaggerated. Earlier that very day, 7 miles was a mere hop, skip, and jump; now, 100 yards was an overwhelming distance to be reckoned with each time one of the five contemplated swimming that stretch.

By the survivors' accounts, one of the women was the first to yield to the prolonged exposure. When she could no longer hold herself to the bow of the boat, she was clutched by the boat's captain, who encouraged the others to remain calm and stay together. On they drifted through the night. At some point, one of the men, seduced by the relative close proximity of an island, attempted to swim, realized that he would not make it, and struggled back to the group. Before reaching the overturned skiff, he floundered and was assisted by the captain, who, in the process of helping, released his grip on the woman who had succumbed. Her body was whisked away, not to be seen again.

It is unclear whether the man attempted to swim again, but he separated from the remaining survivors and disappeared. The second woman was the next casualty, as she quietly acquiesced, putting her face in the water. Before dying, she requested that her body be tied to the boat so as not to be lost at sea. She was secured to the skiff where she floated lifelessly with the final two through a stretch of open water. These two survived. Experts and laymen alike were amazed by this; it was clearly nothing short of a miracle. These two men lasted eight hours in water temperature so cold that any chart would have given them three hours at most.

Islanders don't speak of the incident, certainly the most tragic event in the Island's history, but it has never been forgotten. A plaque in the town hall serves as a memorial to those who died, and is a constant reminder that we are Islanders. To date, I have lost eleven personal friends in what can best be described as six separate showings of the ocean's "conscienceless temper," I have lived through a few of Mother Nature's tantrums when others have not, and these tragic incidents force me to ponder survival. I am often torn between wanting to know more and wishing I did not know as much as I do.

EMERGENCY MEDICINE

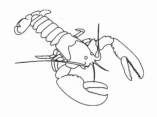

A lot has changed since 1983. We now have emergency medical technicians (EMTs), a few volunteers willing to go through the rigors of training and certification. The EMTs couldn't have done anything for my friends, of course, but were born years later out of legitimate concern over the fact that we have no medical facility on the island. A squad of emergency medical technicians seemed a great solution, perhaps a stopgap measure, but one that had been sorely lacking.

Because the Island community is so small and the town's budget so limited, many tasks appear to go unrewarded. There is always a need for volunteers. I have been told that the definition of a volunteer is "someone who didn't understand the question." If this is valid, there exists a small number of

Islanders who need to pay attention when questions are asked, because they volunteer for everything. Unfortunately, these benign individuals seldom get adequate support, moral or financial, leaving them with feelings of frustration rather than tutelary sainthood.

The best example of an Island volunteer is Theresa Cousins. Theresa is always very generous in offering to get involved in most everything the Island needs. The woman is willing to do anything for anyone. Baking cookies for a sale to benefit the school or putting a quilt together for a fund-raising raffle are among the many things that one can count on Theresa to take care of. These are simple tasks that many could do but don't.

Theresa Cousins, among other things, is a member of both the Island's emergency medical technicians and the budding, struggling volunteer fire department. Although the EMTs have had their share of problems and opposition, they are a well-trained and, some would argue, needed group. Theresa's husband, John, is the head EMT and also served as fire chief, the least coveted position on the Island. John is not apt to bake cookies but is a willing volunteer for any position of authority.

The idea of Island EMTs was first conceived in 1997 by the town's selectmen. Until this time, treatable medical emergencies were few, and were dealt with in a Chinese fire drill fashion. If medical attention was needed in the summer, one of several vacationing doctors was called upon for help. I recall countless occasions during my childhood when reports of accidents or illnesses were prefaced with "Dr. Ellis said . . ." It was unwise to get sick or injured during the winter months, so most people avoided this whenever possible. In cases of extreme emergency, a local fisherman would agree to transport the victim to the closest point of land, which is Stonington. The mailboat had also carried many a victim to medical attention. In dire emergencies, life or death, there are places on the Island where a helicopter could land to expedite the trip to the hospital, but to my knowledge this has never happened.

When I was eight years old, my mother took ill with pneumonia and was rushed off-Island by Billy Barter aboard his boat the *Islander*. And in my teens I recall my cousin Dana doubled up in pain with appendicitis on the deck of the *Danita* while Captain Jack MacDonald drove his lobster boat-turned-ambulance to the dock on the mainland.

Billy and Jack have been relieved by the next generation of emergency boat drivers, mainly Dave Hiltz and me. Me because the *Mattie Belle* is one of the fastest boats in the harbor, and Dave Hiltz because he is willing to venture out in poor weather any time of year.

The closest I have come to needing an EMT myself was one year before their conception. I was hauling lobster traps with my friend Ben MacDonald just west of Kimball Island. Ben and I were enjoying the sun, good fishing, and conversation, when I found myself lying on the *Mattie Belle*'s deck looking up at Ben, who was looking worriedly down at me. Things were fuzzy, but as they came into focus I realized that I had been knocked out. Before I could say "Auntie Em," Ben asked if I was all right. My head hurt. "What hit me?" I asked, sitting up and feeling the contusion that had popped out above my right eyebrow.

"The block," Ben answered, referring to the 4-inch round of steel hanging from the end of the *Mattie Belle*'s davit. The stainless-steel rod whose function was to keep the davit in place was nearly doubled in half. The traps I was hauling at the time of contact had apparently "hung down" (caught on the ocean bottom). The strain of the line between the hauler and the traps buckled the support rod, allowing the block to swing inboard, whacking me on the forehead and taking me off my feet. Ben helped me up and walked me to the stern, where I sat on the transom assessing my injury. I wondered if I should ask Ben to take me to the mainland. Perhaps my skull was cracked, or I had a concussion. My head was pounding. The bump was now the size of a golf ball and throbbed painfully. I thought how lucky I was not to have been hit in the temple, the eye, or mouth. I pulled my hand from the bump over my right eye and inspected my fingertips. "No blood. I guess I must be OK."

Ben helped me to straighten the buckled rod, we left the "hung-down" traps for another time, and continued to the next pair, finishing the day uneventfully. The headache eventually subsided, and I was left sporting a blue-ribbon black eye, which was material for some lighthearted teasing of Ben by many who accused him of "finally straightening the captain out." Looking back, I am certain that if EMTs had been available, I would have had them check me out before going back to work.

It was the general consensus in 1997 that we had been pushing our luck, and the idea of a trained group of EMTs gave many residents peace of mind. Eight interested people entered into an agreement with the town to provide three years of emergency medical service to the Island's residents and visitors in exchange for the Island funding their training, equipment, certification, and licensing. The basic training began in the winter of 1998 and was regarded as a great off-season activity, a time of year when activity is rare. The idea of having something constructive to do during the winter months was both novel and appealing to the year-rounders who enthusiastically signed up.

Of course, there were those "doubting Thomas" types who believed the EMTs were unnecessary at best or even a hazard. Victor Richards, for example, he of the mail-order brides and the Alabama Slammer, had a T-shirt printed with "Do Not Resuscitate!" In retaliation, one of the soon-to-be's had a T-shirt made that said "Don't Fuckin' Worry, Vic!" Most of the community members were happy for and proud of the eight willing participants in the training program, lavishing upon them praise and support. Work schedules were altered, nice people filled in for soon-to-be's, helpful supporters and friends baby-sat dogs and kids while our budding medical techies studied, memorized, and practiced. Eight people is a healthy percentage of our winter population, about a fifth, so just about the entire community was involved in the EMT thing in some capacity.

One hundred and seventeen hours of classroom, and forty hours of hands-on clinical experience later, seven nervous students prepared for the state certification exam. (Only one student had dropped out during the winter.) The state's test had two parts: a three-hour written exam and an intense practical exam. All seven students passed with flying colors. The Island was proud, and anxious to move on to something else.

EMTs had been the topic of nearly every conversation that winter. Naturally, the community's attention span reached its thin end in the spring. It was time to attend to the needs of the summer people. Water was turned on, decks were painted, propane bottles were shifted, and vehicles tuned up. Simultaneously, lobstermen were setting traps and sprucing up boats. Gardens were planted. The number of EMTs dropped from seven to six. But the remaining core was committed and steadfast in advanced training and certifications for upgrades. The public's fading interest was felt most in the town's fizzling funding of EMT projects, training, and equipment. Undaunted, the EMTs began to raise their own funds. A defibrillator was purchased with cash raised through a dance, bake sale, and can for "spare change" next to the cash register at the store.

The EMTs were certified just in time. By the end of our first summer with resident emergency medical technicians, the one noticeable change was in the level of activity requiring emergency medicine. Island Rescue was born, complete with pagers, beepers, and a 911 number. There was a new record set in emergency boat rides to Stonington. Whether this situation was created by an overzealous squad or due solely to coincidence, we will never know. I transported several victims of bicycle crashes and hikers who had fallen on trails, along with the attendant EMTs, to the dock in Stonington where the ambulance and attendants always met us promptly.

I remember my first ambulance run, and can still feel the anxiety I experienced as the captain of the boat. I never volunteered for first-string water taxi (I understood the question), but was happy to oblige. When I answered the phone, Aunt Sally was brief and to the point. She was calling for the

EMTs. Was I willing and available to take one of the masons who was working at a construction site to the hospital? "He's having a heart attack." I agreed, dropped whatever I was doing, and rushed to get the *Mattie Belle* from her mooring to the town landing where I was to meet the patient and EMTs on their first real emergency run.

We had many work crews on the Island that summer, with three new houses under construction. There were masons, Sheetrockers, nail pounders, and two men with a huge hydraulic drill rig for boring holes in ledge for dynamite sticks. I wasn't surprised that a mason would develop a heart problem — all that lifting of stone and mixing of cement by hand in the heat of the day. The definition of EMT service being "prehospital medicine," I was happy to know that ours were certified and available, as the nearest hospital is a minimum of an hour away, thirty minutes of sometimes bumpy water, and thirty minutes of always bumpy road.

As I tied the *Mattie Belle* across the front of the dock, down the ramp came the patient, flanked by two EMTs toting a backpack and an oxygen bottle that was tethered to the mason's face with a fathom of plastic tubing. I was relieved to see the victim come aboard under his own steam. I felt fortunate that it was a clear and calm day for the run to Stonington. The mason's complexion was rather pale, almost gray, and he was sweating profusely. His breathing was somewhat labored, but he seemed otherwise comfortable as I pushed away from the dock and headed for the closest point of mainland. I was nervous. I looked to the EMTs for some sign. They were both busy monitoring the man's vitals, being careful to write everything down in a small notebook. I hoped that if things took a turn for the worse, one of the techies would quietly motion to me to pour on the coal, as I was at a comfortable cruise speed with some throttle in reserve. The term *heart attack* scares me. I felt sort of sick. The EMTs were being so thorough, cautious, and attentive to the mason that they never even gave me a glance. Why would they? He was the victim. They remained calm and confident, so I remained at cruising speed.

About fifteen minutes into the thirty-minute ride someone called on the VHF radio. In answering, I learned that it was the ambulance crew waiting on shore for our arrival, and confirmed that we were fifteen minutes away. When the voice inquired about the nature of the emergency and the status of the patient, I gladly handed the microphone to one of the EMTs, who again impressed me with her command of medical lingo. She answered each question concisely and with confidence.

A quick glance at the patient out of the corner of my eye told me the mason's chest and shoulders were no longer heaving with each breath. He must be feeling better, too, I thought. I assumed that he might like to know that we were almost there, and said, "Five more minutes." When I cocked my head to face him, his color frightened me. He was turning blue, and his eyes were starting to bug out. In a panic, I thrust the throttle up to wide open in order to shorten the trip by one more minute. Both EMTs looked at me with questioning eyes. I nodded my head toward the patient, whom they had abandoned for the radio and who was now looking extremely uncomfortable. The EMTs seemed startled for a second, and then one of them reached with a crooked finger for the oxygen mask that was stuck to the mason's blue face. As she pulled the bottom of the mask, breaking its seal from under his chin, I heard the sound of burping Tupperware. Apparently the oxygen had run out, leaving the mask suction-cupped to his perspiring face like a toilet plunger. The patient breathed deeply. A more natural color returned quickly to his cheeks, and his eyes no longer bulged.

I slowed the *Mattie Belle* to an idle as I entered Stonington Harbor. By the time we tied to the dock, the saucer-sized hickey was no longer discernible around the patient's mouth. I was relieved to watch the ambulance drive up the dock and disappear over the hill destined for the medical center. I got word the next day that the mason was doing fine, thanks to Island Rescue and the water ambulance. The EMTs have since had much practice with heart patients, or with one in particular, who required five emergency runs with heart-related problems in five months. It was suggested by more than one late-night boat driver that this woman might actually have been unknowingly asphyxiating herself by heating with kerosene burners that were not properly vented, and that the EMTs might try the electric paddles on her, but to my knowledge they never did.

On a national average, 25 percent of all licensed EMTs quit their service each year. We are presently down to four EMTs, so in the year 2005 we will be back to square one. One of the EMTs' oxygen bottles was recently spotted being used as a trig behind our fire truck's left front tire. (We don't actually have a fire department or any firemen, but we do have an old truck.) I have heard some speculation that a defibrillator might bring a dead battery to life. If nothing else, Islanders are resourceful, so the equipment will continue to go to good use.

JAMES SPENCER GREENLAW

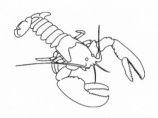

"Dad has a friend?" Bif asked with some degree of skepticism.

"Yeah. I didn't believe it either at first, but some guy is coming to visit. He's a childhood friend of Dad's. They haven't seen each other in almost sixty years."

"Wow. I never imagined that Dad had any friends . . . cool."

By the time my sister Bif and I finished our telephone conversation, we had discussed fully the fact that the only friends we ever knew our father to have were the spouses of our mother's friends. James Spencer Greenlaw, Jim, Dad, never had a buddy. At least not in the forty years that I had known him. He reveals his character through deeds rather than words, so he's hard to get to know in the normal span of time that most people take to form friendships, or to decide whether or not they want to. Dad has never had the gift of gab, and his silence intimidates some people. I think that all I know about my father I learned from watching him. In the years since his retirement, I have had plenty of occasions to watch him, mostly when he's doing projects for my mother in tandem with his partner in crime, his brother-in-law, my uncle Charlie. A typical task attacked by the two progresses as follows:

Dad and Charlie inspect a spot on the outside wall of the house with a keen interest. Charlie leans on a builder's square and watches as Dad pulls out a section of his measuring tape and lays it along the line he has just drawn with the pencil that he then returns to rest behind his right ear. The two men (Dad well into the balding process and Charlie sporting a full head of white hair) face each other and discuss the measurement. Nodding in agreement, Dad again plucks the pencil from his temple and slashes an *X* through the existing horizontal gray line while Charlie examines some papers entitled "Installment Instructions."

"Let's see. Thirty-fourrrr . . ."

". . . and one-sixteenth." Both men now stare off vacantly while silently calculating to themselves. Dad rocks back and forth from heel to toe, and narrows his eyes while he thinks. Charlie stands still, only his lips moving as he counts silently. The two retirees compare answers, agree that something is amiss, and go back to the wall with the tape measure to resolve the discrepancy. The calibrated metal strip again goes against the siding, but, important difference, this time vertically from the ground up. The phone rings inside the house. Their concentration on the project keeps the men from hearing both the phone and my mother's conversation.

"Hello."

"Hi, Mom. It's Linda. Have they cut the hole yet?"

"No. They're *still* measuring." My mom then describes the all-too-familiar scene to me. The men seem to hear that all right, and share a knowing look that confirms a conviction they have held for decades of marriage: Women just do not understand the importance of accuracy in projects such as this.

"Jesus, good thing they're not charging you by the hour." My mother invites me to join them for lunch, warning me that the bread may have become stale waiting for the "engineers" to take a break from the latest project she had devised for my father, and that my uncle Charlie had been kind enough to offer some help with. Actually, Charlie's assistance may have been volunteered by Aunt Sally, who always seems as anxious to get Charlie out of her hair as Mom is my father. Retirement on the Island is not sitting around and enjoying the golden years. It is as much work as full employment; the women see to that, often volunteering their husbands' services to anyone with an odd job.

Charles Bowen, my uncle Charlie, is a retired machinist. Years of measuring, cutting, and grinding steel to within a tolerance of one-ten-thousandth of an inch have honed his personality and mannerisms, which are nothing less than precise. Whether he is telling a story or building a sandwich, Charlie's beauty is in his attention to the smallest of details. Dad, too, has impeccable manners, and is quite deliberate and punctilious in all aspects. My father is so methodical in both word and deed that he would surely drive the average amateur home repairman mad with frustration. The two old men work well together because they share an overzealous enthusiasm for being neither overzealous nor enthusiastic. Dad and Charlie take "measure twice, cut once" and multiply it by a factor that is astronomical.

I, on the other hand, am the prime example of the roughshod slam-banger, having learned a great deal of impatience from my best and oldest friend, Alden Leeman. (Readers of my first book, *The Hungry Ocean*, may remember Alden, the hot-tempered captain who taught me almost all I know.) Had Alden and I taken on the chore of installing a new stove in my folks' house, we would have cut three or four holes through the wall for the vent by now, and all of them would have been wrong. (Not just wrong, but way off.) Where my father required a jigsaw, Alden would need a chain saw; my father, sandpaper, Alden, an ax. After having spent so many years working for Alden, and just now getting reacquainted with my father, I can't help but compare and contrast the two most important men in my life.

Relationships with male friends, fathers, and brothers are weightier when there is an absence of husbands and lovers in a woman's life. Alden's opinion of my sustained singleness is that I intimidate many men, so they do not get to know me well enough to truly dislike me. If Alden's theory has merit, then it becomes a bit of a vicious circle because I am not interested in men who are so easily intimidated. Dad, as opposed to Alden, who is constantly advising me in personal matters, keeps his opinion to himself.

I have spent much time waiting for Mr. Right, who does not appear to be looking for me. (My older sister, Rhonda, once referred to *The Hungry Ocean* as a book-length personal ad, and will no doubt view this as another one.) I was now at a point of not quite asking for help, but certainly willing to listen to what was unsolicited in how to "trap" a man. Years of longlining for swordfish had me thinking that baiting, hooking, and hauling was the process by which a man would be caught. But since coming ashore to fish for lobster, "trapping" had become the preferred metaphor, and it was proving equally unsuccessful.

Coming to the Island to start a family was a goal I had set for myself, and one of the main catalysts for my decision to give up offshore work. I had some very nice boyfriends during my swordfishing years, but there's just something about "Thanks for dinner. See you in thirty days" that is not conducive to second dates. Since coming ashore, my status had not changed, forcing me to consider the possibility that my fishing schedule had not been the problem. Moving to the Island was not the best way to facilitate the family plan. There are three single men in residence; two of them are gay and the third is my cousin. To make matters worse, I was finding lobstering tougher than I had expected, especially this season, where we continued to change the water and the lobsters just weren't coming calling. It didn't take me long to realize that the only thing I do like about lobstering is working with my father. Dad is good company because he really isn't any company at all. He works. People often misjudge my father as being too serious or stern. Once the same people get to know him, they learn that although he was not blessed with the gift of gab, he has a wonderful sense of humor. Like a good martini, it is very dry.

For example, a couple of months ago, Dad and I were in the patch of woods where I stored lobster traps for the winter, working to prepare a load of traps for the water. In the midst of this work, we became aware of a man and woman peering from the road through the trees at us. "Do you know them?" I whispered to Dad.

"No. Must be day-trippers," Dad replied, referring to visitors who arrive on the "early boat" and leave on the same day's "late boat." As we continued to overhaul the pile of traps, the couple edged closer, seemingly fascinated by the work that we found absolutely humdrum.

The woman peeking through the brush cupped a hand to her mouth and yelled, "Hello in there!" Because we were hustling to get thirty traps out of the woods, into my truck, and onto my boat before the tide started ebbing, we gave an abrupt greeting to the strange couple and kept our heads down and hands busy to discourage any conversation. The woman again tried to engage us with a question. I was sure that I must have misunderstood her or missed something.

"Did she ask what I think she asked?" I quizzed my father. He smiled, shrugged, and shook his head, indicating to me that he had not quite understood either. The couple waited for a reply.

The strangers then took a few steps off the road and into the woods, closing the distance between us slightly. The woman yelled again. This time the question came through loud and clear. "Are you catching any lobster in those cages?" How could I politely answer such a stupid question without getting into a lengthy discussion, including some very basic facts about lobsters? (Like, they don't live in trees.)

"Did you hear her that time, Dad?"

Again my father shrugged and smiled, then replied to the couple: "This hasn't turned out to be a very good spot. This may sound crazy, but we're going to try moving some of these cages into the water today. Maybe that will be better."

"Oh well, good luck!" The woman was visibly pleased that she had spoken with a genuine Maine lobsterman, and she and her husband continued their stroll.

My father and I have fished together ever since I can remember. I was, in a way, my father's only son until my brother Charlie was born. And even after his birth in 1968, I stayed on as our dad's fishing (and hunting) buddy until Charlie, who eventually became Chuck, grew old enough to tag along and finally become the son when I outgrew the position. I was happy to have it again now that I was back and Chuck and his family were on the mainland. But it took Dad and me a while to regain ease in working together. Dad had learned that if he ignored 90 percent of what I had to say aboard the boat, we were both much happier. And 90 percent of what I have to say while hauling lobster traps is complaining about hauling lobster traps.

Each morning, as I climbed over the rail

of the *Mattie Belle* to begin a day, I felt I had ample cause to complain. The lack of lobsters began causing my disposition to sink to uncharted depths. My friend and former captain Alden would have given me grief about my mood, and we'd have ended up fighting. Dad just went about his onboard chores. Daydreaming became my escape. When I wasn't bemoaning my fate, I was imagining myself in another life. I fantasized about going back to offshore fishing. After all, my scheme to satiate my nesting instinct had failed. The admission that my plan to start a family had not been thoroughly thought out was taking a toll. I felt quite badly that I was getting increasingly hard to get along with, but couldn't seem to do anything about it. I could barely stand to be around myself. Unhappiness was something I had known so little of that I scarcely knew how to begin dealing with it. I had always been confident that I could achieve anything through hard work; the difficulty was only in deciding what I wanted to do. I had spent most of my life happily transiting from point A to point B to point C. . . . At this present stage, deciding on the next point of destination was agony. Maybe I would go to Alaska. . . .

As the month progressed and August

lurked, my daydreaming started to make me careless. One day, reaching into a trap, I felt a sharp twinge of pain race up my arm. In a jerk reflex, I pulled my hand out of the lobster trap to see what was causing the harrowing pain that started at the tip of my thumb. A large crab had clamped onto my finger and now dangled by a claw from my gloved right hand. "Ouch! You son of a bitch!" I flung both hand and crab toward the deck; the angry crustacean released its hold on contact. "Take that, you motherfucker!" I stomped the crab with the heel of my boot repeatedly, until it looked like it had been run over by a bus; I continued swearing with every stomp and grind. I had managed to flatten the shells and guts into a mass twice the crab's original diameter when I became acutely aware of my father's presence beside me.

My father seldom curses, and although he has never scolded me for it, I know he disapproves of the language I use aboard the *Mattie Belle*. I followed my father's stare from the dead crab to eye contact to the crab again and back to eye contact. "Wow, that hurt," I said.

Dad scowled and focused again on the dead crab. "I'll bet it did," he said quietly. It was much less from him than I wanted. In a move to gain sympathy, I pulled the glove off to inspect the thumb that was now pulsating between excruciating pain and dull ache. The nail was already purple and blood trickled from the cuticle. I assumed that my father would be concerned enough to examine the damage or at least indulge me with some words of condolence. But he was not and did not. I lowered my expectations to wishing that he would give me the satisfaction of telling me that I was a horse's ass. At least then I could argue with him and complain that *he* had not been the one bitten. But he did no such thing. Dad turned back to rebaiting the traps that rested on the rail, seemingly oblivious to the pain and humiliation I was suffering.

In one final act of frustration, I gave the Frisbeed crab a sound kick, sending it to the stern like a hockey puck. "Bastard!" I yelled, not knowing whether I referred to the crab or my father.

I had never hated lobstering more. Running the boat harder than normal, hauling faster than usual, and slamming things around in general, my anger slowly dissipated to self-pity with the hauling and setting of two strings of gear. Oddly, I soon became aware that I was gaily singing "King of the Road," and imagined my father might suspect me of being schizophrenic. Day-dreaming always succeeds in pulling me out of self-induced emotional funks, and if I am singing, humming, or whistling, I am content within my own world. When I catch myself in the midst of a tune, I am never sure how many times I've repeated the same verse, but have been told by many a crewmember my repetitive singing is incredibly annoying. (I was once haunted by the chorus of "The Cover of the Rolling Stone" for a period of two weeks.) The singing I get from my mother. I knew every word to Petula Clark's "Downtown" well before I learned "Twinkle, Twinkle." Mom sings constantly when she's in the kitchen, which may be my connection between song and happiness, since my mother's cooking always makes me happy. "King of the Road" is still one of my old standbys; I was relieved to realize I was singing it while trying to escape the pain in my thumb and the reality of hauling largely empty traps.

In the midst of the third verse, I paused to yank a trap onto the rail and slide it aft to Dad, who waited for it. Another round of the chorus, and the second trap now rested in front of me. I opened the door and was delighted to find a feisty 2-pound lobster. As I eased the lobster from the trap, I was disappointed to see that it was an egg-bearing female (egger) — back to the ocean with her. I turned with egger in hand toward my father, who was slipping by me with a freshly baited bag for my trap. A familiar voice on the VHF radio distracted me, and in an instant the egger had latched on to Dad's bare forearm with both claws. A lobster has two different and distinct claws — a "crusher" and a "ripper" — aptly named for their two separate functions, which they now performed on the white, sensitive underside of my father's arm, midway between wrist and elbow. She had both claws full of flesh and showed no sign of letting go.

I jumped out of the way to give my father room enough to swing the lobster against a bulkhead or splat her onto the deck as I had the crab. Dad stood patiently, supporting the weight of the lobster that hung on his left arm with his right hand. He was gritting his teeth, and he squinted as he waited for the lobster to relax and release. "Jesus, Dad. I'm sorry. She grabbed you so quickly. I wasn't paying attention." I felt terrible.

"Do you think you can gently pry her claws open?"

"Lay her on the rail." I was eager to help ease the pain I had caused. "I'll smash her for you." This offer was rejected with a disgusted sigh and a shake of his head. Pregnant female lobsters are always the most aggressive, but this one eventually released her ripper. I suspect she was thinking about getting a better grip, but before she could, Dad had the freed claw in his right hand, where he held it closed until she let go of him with her crusher. I tried to look at my father's arm, but he ignored me as he gently tossed the lobster into the water, being careful to land her on her back so as not to disturb the eggs on her belly.

"Bitchy female" was all Dad said. And as he returned to the bait bags, I wondered whether he was referring to the lobster or me. Assuming that "bitchy female" was referring only to the lobster, I defended her disposition, siding with the lobster in a male/female analysis that occupied my mind while my body performed what was necessary to finish the day's work. I had done a fair amount of reading pertaining to the biology of lobsters, and knowing that a female's gestation period can endure for up to twenty months, figured the egger had every reason to be bitchy.

As far as I can see, there is absolutely nothing attractive about a lobster, male or female, and I have often wondered how so many manage to reproduce. Obviously, lobsters find one another attractive enough to be remarkably prolific, although the mating rituals seem more perfunctory than romantic.

When the female lobster molts or sheds (loses her shell), she is as vulnerable as any girl who disrobes. Molting is triggered by water temperature and generally occurs around mid-July in the area I fish. (When I hear a radio report of "shedders in Duck Harbor," I can't be sure whether the fisherman has caught a soft-shelled lobster or spotted nude sunbathers on the shore.) The female lobster appears to dictate when the mating process will occur (not unlike humans, or so I'm told), for prior to molting she seduces the male out of his den with a squirt of pheromone that acts as a sexually stimulating perfume. The perfume, once emitted, attracts all male lobsters in the vicinity.

Levels of intelligence and bank accounts are not the criteria by which the female lobster judges suitable husbands. A lobster's prerequisites for paternity are simple and all physical. (How anyone has been able to determine this is beyond me. But this all seems to be common knowledge among those who have nothing better to do with their lives than study lobster sexuality.) From among the males she has enticed out of hiding, the female selects the one that is the largest and strongest. The exception being that Mother Nature does not permit lobsters to mate with siblings or first cousins. (No, not even in Maine.) Once she sets her sights on the stud, the female persuades him by sending his way a stream of urine loaded with pheromones. Intoxicated, the male advances aggressively with claws up. Depending on the nature of the particular female, she will either spar or turn away. Neither an overabundance of willingness nor being coy discourages the male, who is now quite persistent.

Eventually, the male lobster coaxes the female into his den, where they may remain for up to two weeks, waiting for the seductress to shed her shell. Once out of her shell, the female is in a state of precarious vulnerability, and the male must choose between two options: mate or eat. Voyeurs confirm that although the male may be quite hungry by now, he *usually* opts to mate with rather than consume his date. (I suspect that choosing between eating lobster and having sex would be more difficult for the human male.) The act itself, the experts report, is performed with "surprising tenderness."

After mating, the female hangs around the den only until her new shell begins to harden a bit. Once protected with the new shell, she leaves her beau without so much as a backward glance.

While mating, sperm is deposited into a receptacle within the female's body that acts as a sperm bank. The female has the ability to store sperm for months, until her personal agenda permits pregnancy, at which time she fertilizes and lays her eggs. Lying on her back and cupping her tail, mother lobster pushes from her ovaries up to twenty thousand eggs that are fertilized as they leave the storage compartment. The fertilized eggs adhere to the underside of the female's tail, where they ride for nine to eleven months. The doting mother fans her eggs to keep them oxygenated and clean until the time comes to release them into the ocean, where they drift aimlessly with the current. Giving birth to twenty thousand eggs is quite a task and can take up to fourteen days.

It is estimated that, due mainly to predators, only ten of the twenty thousand released eggs will survive long enough to resemble what we think of as lobsters. After hatching from the egg, the lobster progresses through four larval metamorphic stages, developing into the strange creature that lobstermen refer to as a "bug." Drifting within a meter of the ocean's surface, larvae are easy prey for seabirds and fish until they drop to the ocean's floor in their fourth stage, between two and four weeks of age. Baby lobsters molt into a fifth stage, this time finding a place to hide for a period of up to four years. (It is tempting to suggest that certain children should try this.) During the time of hiding, lobsters venture out only to feed, and this only at night. Lobsters are nocturnal.

A lobster molts (which is how it grows) up to twenty-five times in its first five years of life. A newly molted lobster is somewhat fragile and extremely lethargic, an easy target for anything looking for a meal. Once an adult, the lobster molts once a year until it achieves maximum growth and maturity. A legal-sized and marketable lobster is approximately seven years old.

It seems quite clear that a female lobster spends the bulk of her life pregnant, molting, or hiding from predators. With a life like that, who wouldn't be a bitch? I wondered how the plight of the lobster could be stretched to somehow justify my own irritability of late, but abandoned the ridiculous effort as we called it a day and headed for shore. Once ashore, I yelled to Dad that I would be home in time for dinner; he smiled and waved as he always does. His forearm was a mess where the lobster had grabbed him. I was aware that although I had whined about my thumb several times that day, my father hadn't mentioned his arm.

GEOLOGY

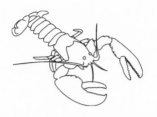

The threat of depopulation and its effect on a small, fragile community can be seen as somewhat of a self-fulfilling prophecy. Single folks fear they will have few choices of mates, and leave — so new, young families, which are the foundation of any community, become almost nonexistent. There are many barometers of a shrinking populace, but one of the more tangible ones on the Island is the increasingly shorter hours during which our general store is open for business.

The store is exactly what one might expect from the only shop selling groceries (or anything else for that matter) on a small island. I expected that it was my best chance for tracking down the Island Boys, George and Tommy, as they rarely missed having lunch there, joining a few of their cronies at noontime daily. The structure itself is a saltbox of a nondescript color; it is either gray or white, I'm not sure. Clapboards are missing in a few spots, and the sign over the lopsided porch is faded to a degree that might suggest the building had been deserted. A broken screen door flaps in the breeze beneath the eaves, where house sparrows nest, offering fresh droppings as evidence of life. An off-blue picnic table sits between the store and the fuel dock, which juts out into muddy clam flats at low tide. The store's single self-serve gas pump, which has not and may never be upgraded to digital, has been backed into so many times that a stone barrier was built around its base for protection. The stone wall appears to have been nudged by more than a few fenders.

I wasn't sure the Island Boys were intentionally avoiding me. But taking no chance of being detected, I walked to the store, leaving my pickup in my aunt Gracie's yard. Word travels fast with so few ears to receive it, so by now the Island Boys were on the lookout for the faded maroon "battle wagon." With no exhaust or muffler, the truck did not allow for a sneak attack.

George and Tommy are the quintessential Island suckers. They suck all they can from year-rounders and summer folks alike, and add nothing other than a bit of entertainment. When the two men first arrived on the Island, they formed a business partnership called Island Boy Repairs, hence their nickname. Nearly everyone has fallen victim to the team at one time or another. What most irks the locals is their constant "that's not the way we did it on the mainland," which quite naturally leaves us all wondering why they do not return there. But because of the shortage of willing and available employees, George and Tommy manage to stay quite busy. What really sickens me is the fact that I was their most recent victim. And they had yet to repair the job they had blundered on my behalf. So that's what I was doing at the general store, trying to track them down to get them to finish what I didn't need done in the first place. I should have run and hidden when I had heard their truck approaching the week before.

The sound of their truck had come from behind me as I stepped from the road into the trap lot. I had turned to see their dilapidated Ford roll to a stop. Behind the cracked windshield sat the two big, burly men, my old buddies George and Tommy. Printed down the side of the truck in sloppy white paint is

ISLAND BOYS REPAIR SERVICE — IF WE CAN'T FIX IT, IT AIN'T BROKE. There follows a list of the jobs for which the men fancy themselves qualified: "Automotive work, home maintenance and winterization, landscaping, small engine repair, fiberglass construction, painting, caretaking."

Tommy, a self-proclaimed ladies' man, appeared not to have bathed since the pond had frozen over last fall. He threw himself against the inside of the door; it's a sticky door desperately in need of repair. (As a side note, Tommy subscribes to the theory that if he propositions every woman he meets, eventually one will say yes. So far I was sure that had not happened. Through the years that I have known him, I have been honest with Tommy about my opinion of why he hears "no" so often; he is absolutely filthy at all times.) George slid across the truck's seat and climbed out Tommy's door, explaining to me, "Driver's side door doesn't work."

I was at that time glad to see the two men. George and Tommy had moved to the Island from Baltimore and Philadelphia, respectively, twenty years ago. Although I suspected that they were running from the law, it was rumored that both men had left top management positions — George in insurance, Tommy in the medical field — to come to the Island on vacation, and had never left. They became enthralled with the laid-back Island way of life, and simply refused to reenter the rat race at the end of what was intended as a two-week holiday. Their wives, who preferred city life, divorced the men, settling for nothing less than everything, which for George was several million dollars, and was in his own words "the best money I ever spent."

As we talked, George surveyed the lot around us and finally said, "Some of these trees need to be taken down before they get blown over. Tommy and I can do that for you. I have a new chain saw."

"Well, thanks, but I can't afford to hire you until I start catching some lobsters. Maybe later in the season."

"It won't cost you much. And you can pay us when you start making some money. Come on, Tom. Let's go get the saw. We've got some fuckin' wood to cut!" And into the truck they scrambled before I could say "no."

"That truck isn't a very good advertisement for your handiwork," I remarked.

"Are you kidding?" Tommy spoke through the window as George tried to start the engine that coughed in refusal. "This truck is testimony to our genius. The fuckin' thing is fifty years old. We are miracle workers, wizards." The engine finally caught and roared. "See?" They faded down the road in a cloud of dust, and I hoped they would get sidetracked before returning with the chain saw. In the meantime, Dad had showed up to help cut and splice rope that we would need in the fall to move traps to deeper water.

Dad and I worked in the shop, using the electric rope burner to cut the line into different lengths, splicing an eye into one end of each section. We estimated and calculated what number of what length pieces would be needed, and spliced until my fingers were sore. The deafening sound of a chain saw at close range made it impossible for Dad and me to continue our figuring and planning. I didn't even look up at my father, but kept my focus on the line I was now splicing, thankful for the noise that prevented him from asking, "Did you hire those idiots?"

A loud crash then shook the shop. It was followed by a short silence and then a clearly spoken "Oh, fuck." Dad and I rushed to the door, from there we could see the tip of the giant evergreen tree that had been felled right onto the roof of the building in which we stood. The toppled spruce rested in the dented peak like an overgrown baby in a cradle. Around the corner of the shop ran Tommy and George in hard hats and safety glasses. The armpits of Tommy's T-shirt were sweat-stained all the way down to the love handles that jiggled above his belt. George ignored Dad and me, and said, "Come on, Tom. Let's go get a ladder. We've got a fuckin' roof to fix."

As I stared dumbfounded at the damaged roof, the two men vanished. I heard the squeaking open and banging shut of the truck door, followed by the sound of a sick engine that slowly perished into silence. I was relieved that my father chose not to discuss the incident, and hopeful that the Island Boys could replace the rafters and shingles before the weather changed to rain. Knowing what I did of the diligence of the men, I had visions of them lending me buckets with which to catch the streams that would surely run onto the expensive rods and reels I tried to protect from the weather.

One week had passed since the roof damage, and I had not laid eyes on the Island Boys. Dad and I had removed the tree, and I had been praying for dry weather. I wondered how much the Island Boys would charge me to repair the damage they had caused. By the looks of the lobster barrels I'd been filling, they would need to be awfully patient to collect any money from me. Trap after endless trap over the last week had come up empty. The season was looking to be a bust.

I entered the store at twelve o'clock sharp hoping that the Island Boys were there. They weren't. I greeted my aunt Sally, who manages the venture, and my longtime friend Ben MacDonald, who's the store's employee. One of the Island's handful of remaining lifelong residents, Ben moved here at the age of three days and stayed with his grandparents after his father was lost at sea. Other than two stints of off-Island schooling, Ben has never lived anywhere else and has not as of the time of this writing traveled farther from home than Maryland. Ben minds his own business, which enables him to remain friendly with all parties on either side of the many rifts and squabbles endemic on the Island. The ability and desire ultimately to "get along" is what makes Ben the pinnacle of Islandness.

Ben left the Island for the first time in 1972 to attend high school, and returned to the nest four years later with the knowledge that he is gay. By the time we had reached adolescence, I suspected that my buddy might be gay, but apparently Ben knew nothing of the label until he joined the freshman class at the public high school in Vinal Haven, where the term *faggot* was defined for him. It is no wonder Ben did not realize his orientation until leaving home.

The only student enrolled in the Island elementary school for a three-year period at a time when the population really looked to be petering out altogether, Ben was eventually joined by three others, two of whom were his cousins. For those of you who don't think anyone is born gay, Ben is proof you are. He certainly never had a gay male role model from whom to learn. I am aware of the common belief that gay men make great friends for women because "sex never threatens the relationship." But that doesn't even begin to describe why we are friends. Part of the reason, I guess, is our joint membership in the "lonely hearts club," an association shared by all unattached residents, gay or straight. I also like Ben and am comfortable with him, two things often lacking in love relations. "Like" has become a stronger and more meaningful word in my vocabulary than "love." To truly *like* someone is deep. Ben and I were in agreement that it would indeed be nice to have more people in residence to like.

"Hi, Ben! Hi, Aunt Sally!" I smiled as I closed the door behind me. "Have you seen George and Tommy?"

"No. They haven't been in yet. They're down on the schoolhouse shore hauling Victor's boat out of the water to paint it," answered my aunt as she prepared the coffeemaker for the lunch crowd that had yet to appear. "The nicest people came over on the mailboat this morning," she continued, "two couples from Martha's Vineyard, hiking for the day. They would love to meet you. They'll be here soon."

"Oh?" I wondered what my aunt had told her new acquaintances to make them anxious to meet me. She wasn't about to say, but she did grin, which made me a bit nervous. There is no mistaking Aunt Sally for anyone other than a Greenlaw. She is the female version of her four brothers, the youngest of whom is my father. The most prominent feature of her lineage is the brown eyes, just slightly bigger and much darker than most people's. Even at the age of seventy-five, Sally's eyes retain a mischievous twinkle that reminds me of her talent to walk the fine line between fun and trouble like no one else.

"I can't believe Victor trusts the Island Boys to haul his boat," I said, fishing for some confirmation of my attitude toward the pair, but getting none. I picked a paperback booklet from the small display next to the bread rack, *The Geology of Isle au Haut, Maine* by Marshall Chapman, and sat down on a red plastic milk crate to read the work of one of the Island's many interesting part-time summer residents.

Nearly an hour passed as I skimmed the pages of the geological survey of the Island. Several customers came and went, purchasing staple items. Most of the conversation included the overengineered cradle built by Tommy and George with which they planned to haul Victor's boat from the water and onto dry land, sled-fashion, behind their truck. After hearing some of the structural details of the cradle, I was relatively certain that the lumber that I had been billed for to fix my roof, which had not yet appeared in my yard, had been used in the construction of the boat-hauling sled. This sort of creative billing and borrowing occurs all the time on the Island, as resources and raw materials are hard to come by.

During the winter months, propane gas bottles and tires are the most sought after items and are often borrowed from absent summer people. One-hundred-pound propane tubes are disconnected from houses, and tires are taken from parked vehicles that are left jacked-up until the tires are returned minutes before the unsuspecting summer folks disembark for their vacations. No harm, no foul.

Because my reading of Marshall's booklet had been peppered by interruptions from Island people entering and exiting the tiny store, when I reached the final page, I decided the book could have appropriately been titled *The Genealogy* (instead of Geology) *of the Island*. Endless comparisons between the formation of the physical Island and the fabric of its population came to mind. History does indeed repeat itself. Three geological events that formed and shaped the Island over the last 420 million years have been reiterated in relatively recent decades, resulting in the diverse group of individuals who are all proud to claim the status of "Islander."

The foundation of the Island is bedrock, formed by the eruption of volcanoes. The first settlers were somewhat volatile, as their main objective was to separate themselves legally from the mainland by petitioning the Commonwealth of Massachusetts for a grant of the Island. Episodes of volcanic activity, flowing and cooling, and flowing and cooling, produced layers of granite and diorite. Groups of people flowed and settled, flowed and settled, until several distinct layers, generations and communities, were thriving. In 1878 there were four school districts on the Island, with an elementary school enrollment of eighty-two students. In its heyday, the Island boasted a population large enough to support several stores, the church was erected, and a lobster-canning factory was in full operation. With such close proximity to the productive fishing grounds, the Island's main thoroughfare was bustling with boat activity. Then something changed — a glacial period.

Marshall had written of "the Wisconsinan Epoch," with ice 2 miles thick. "The base of a glacier is a very dirty thing. Although the ice itself is certainly softer than the rock upon which it lies, the ice is loaded with rocks and boulders, which act like sandpaper on the rock underneath it. The glacier exerts a tremendous force on the rock. Glacial movement left gouges in the bedrock. Melting water collects in cracks, refreezes, expands, breaks rocks apart. Advancing glacier plucks broken pieces, leaving bedrock sharp and craggy."

In my very forced mental analogy, the ice age that shaped the population had several similar features. A wealthy summer colony emerged around what began as an exclusive bachelors' club, bringing with it regular ferry service to the mainland. The invention of the gasoline engine meant fishermen didn't need to live closer to their grounds, so many moved inland, off-Island. In the early 1900s, high school education became mandatory, forcing families with teenaged children to leave the Island for Deer Isle, where the nearest high school was located. In 1944, the secretary of the interior accepted, on behalf of the federal government, a gift of nearly 50 percent of the Island, which became Acadia National Park. This conglomerate of glacial activity certainly left the bedrock "sharp and craggy."

In my mind, the Island's most interesting feature is the prominence of "glacial erratics," or "boulders laid down willy-nilly as the glacier melts." There is a certain percentage of the Island's population that I would consider simply deposited here or left behind. One of the clearest examples is Ed, a treasure left on the beach by the tide. Bearded and bandanna-ed, Ed appeared in the thoroughfare in the 1970s with a woman and two small children aboard a demasted sailboat that they called home. Two children soon became four. The woman left years ago. The children are all busy with school and jobs, and Ed remains exactly as he appeared twenty years ago, unkempt and disheveled. Ed looks like he may have been evicted from skid row, but is a constant reminder not to judge a book by its cover. Ed eludes most characterization. A Renaissance man, fantastically well read, sporadically articulate, antiestablishment, an accomplished handyman, extreme non-conformist, Ed is my favorite weirdo.

The last time I had seen Ed, he had been lying facedown in the middle of the dance floor at the town hall, where he appeared to have passed-out mid-boogie. Today, Ed interrupted my reading and we struck up a conversation about the dance and the number of people who had overindulged. "I don't get nearly as drunk as I used to," Ed commented. Ben and I shared a look.

"Really?" I asked. "I wonder how drunk you *used* to get, because I saw you do a swan dive Friday night."

"Oh, well . . ." Ed explained with a sheepish grin. "I used to fall hard and wake up the next day with sore nose or a loose tooth. But now I feel it coming and eeease myself to the floor gently."

"Well, I'm glad you have it under control!" Ben exclaimed. Aunt Sally ignored us.

Ed made his way around the counter and to the coffeepot, where he poured himself a cup. As he added cream, he stated that he was busy helping Lincoln Tully install a new hydraulic motor on his boat's pot hauler. After his usual lunch of two cans of sardines, into which he squirted some mustard from the store's cooler, Ed paid and left to finish the job with Lincoln.

Remembering my search, I paid Ben for the booklet and exited the store to keep looking for George and Tommy. But as soon as I set foot in the tiny parking lot, I ran into a crowd of four people — the strangers I assumed to be the folks from Martha's Vineyard whom Aunt Sally had mentioned. I said "Hello" to the two couples, and one of the women asked, "Are you Linda?" I answered in the affirmative, and the five of us pieced together quite a pleasant conversation of the weather, islands, and lobster fishing.

When the more talkative of the two women caught me admiring her cap (a tan baseball style with "Menemsha Blues" and the outline of a fish embroidered in cobalt thread), she pulled it from her head and handed it to me. I slipped my ponytail through the hole in the back above the adjustment strap, and pushed the visor down over my forehead. It just fit. I liked this woman. It seemed odd that she would, without a word, give me the hat from her head, and odder still that I would not politely decline the gift. I couldn't help but smile and say, "I like your watch," glancing at the gold band around her wrist.

Herb, the quietest of the four visitors, was, I was surprised to learn, a lobster fisherman. Although I had never been there, Martha's Vineyard invoked an image that did not include rubber boots, dirty boats, and stinking bait. I guess I had always thought of the Vineyard as more of a haven for sail-boaters and pleasure fishermen. Herb appeared to be a bit uncomfortable with this outing. I began to wonder what its purpose was. These four people did not look like the typical "day-trippers." They had no bikes, no backpacks. I didn't smell any bug repellent, so they had no intention of hiking any of the many trails that wind around and through the park. I was beyond suspicious when one of the women blurted out, "We have a friend at home you must meet."

"Oh?" I replied, as uninterested as I could manage to sound. I suddenly realized what my aunt had been up to.

"Yes, a guy friend."

"Really?"

"We've known him forever. He's very nice,

and he's single, and he's a fisherman, too!"

Herb, who hadn't spoken a word up to this point, nearly choked, "Fisherman?"

"Well, yes. He has a *charter fishing* business," the woman clarified.

Oh wow, I thought. A Charter Boy from the Vineyard — I'm sure he would be impressed. He's probably an avid catch-and-releaser. Maybe he's even a member of Greenpeace. Ours would be a match made in heaven. Cupid had outdone herself this time. I'm here, he's there . . . What are the chances that we would ever meet? Aren't there any single women on *his* island? Remembering my manners, I exclaimed, "I would love to meet him sometime." He probably uses those fish-friendly hooks with no barbs. I'll bet his boat has a teak deck. Fisherman, ha! "Nice to meet you. Enjoy the Island. See ya." And off I went toward the schoolhouse shore, continuing the search for the Island Boys.

When I passed the town hall, though, I was still thinking about Charter Boy, and laughing about what I imagined his reaction might be to his friends' description of me. Certain adjectives would no doubt come to his cultured mind when told about this female commercial fisherman on this rinky-dink island up in Maine. Barbaric, course, uncouth, rude, illiterate . . . a product of generations of incest with no teeth. My reputation for slaughtering the entire swordfish population in the North Atlantic single-handedly would cause him to choke on his latte, perhaps soiling the front of his silk shirt. He would certainly have no interest in meeting me, unless it was just for curiosity, like going to see Wolf Boy at the circus freak show. Well, I had clearly decided that Charter Boy must be quite an aristocrat — a snob — arrogant, pompous, ostentatious, pedantic . . . I knew a few adjectives myself, and he could go to hell.

I neared the schoolhouse and was again reminded of the toll of depopulation. Seven students this past year, down from eleven the year before. There aren't many one-room schoolhouses still functioning in the United States. The Island school is one of only fourteen one-room schoolhouses in the state of Maine remaining on outposts surrounded by salt water.

Passing between the school and its swing set, I heard the sound of an engine laboring. As I crested the hill that had obscured my view of the shore, I stopped in my tracks to decipher the confusing picture before me. A long, gradually sloping field that stood between the water and me was cut in half lengthwise by what I knew was an operation commanded by none other than Island Boys Repairs. The town's archaic fire truck was parked about midfield, facing the shore. The truck's back bumper was tied to the base of a large tree with a 50-foot length of heavy-duty rope. The spool of the hydraulic winch mounted in front of the truck's grill was bare except for two wraps of wire remaining after the rest had been paid out, and the far end was attached to what appeared to be the new cradle that was partially submerged in the incoming tide. The engine I heard belonged to Victor's boat, *Pair of Aces*, which was trying to force itself into the cradle that clearly had not been placed far enough down the beach at low tide. (Victor had not yet ordered his catalog bride, and the Alabama Slammer had left him. The only woman in his life at this juncture was his boat, the old and faithful *Pair of Aces*.)

George and Tommy paced at the edge of the water. Checking their watches every few seconds, the men argued about how many minutes remained until official high tide, and whether they should abort this mission, push the cradle down farther at ebb tide, and try again when the tide recovered in twelve hours. Victor, who is notoriously low-key, pushed the throttle of his 27-foot Novi-style lobsterboat up a bit more, hoping to gain the last couple of feet needed. He tried reverse in an attempt to back out of the cradle, but it was no use. The boat was stuck. The captain shrugged and chewed the end of his cigar.

"Hi!" I startled George and Tommy, who were concentrating too hard on the problem at hand to be embarrassed or unnerved by my appearance. "Are you trying to pull the cradle up or the tree down?" I asked.

"Very funny. When's high tide?" George inquired. I looked at the rocks and clumps of old seaweed that outlined the high-water mark, and answered honestly that I thought they had at least another hour of incoming tide, and that the additional water would certainly be enough to float *Pair of Aces* fully into the cradle.

Victor, who had idled his engine down and out of gear, yelled from the helm, "Hey, Linda! Looks like I might be stuck here awhile. Why don't you take my truck and grab a six-pack of beer from my fridge?" Thinking Vic's suggestion a good one, I took off in his blue Chevy.

When I returned with the beer, George and Tommy were still fidgeting nervously, and Victor sat on the rail of his boat, his usual tranquil self. As I tossed Vic a bottle of beer from the edge of the water, I thought the confidence he had in the Island Boys' ability was alarmingly misplaced. I wanted badly to mention my roof, but decided that I should not distract or upset the men until *Pair of Aces* was successfully hauled, high and dry. Perhaps I should change my opinion of George and Tommy, I thought to myself. They were certainly taking this job seriously, and I knew their intentions were good. Anyone can make a mistake. As soon as this chore was done, I would insist upon having my roof placed next on their list of priorities.

Tommy helped himself to a beer, stretched out beside me in the tall grass, and gulped down a couple of swigs. Wiping his mouth on a filthy sleeve, he belched loudly and moved closer to me, leaning toward my ear, as if to tell a secret. Tommy softly sang a line from a familiar Jimmy Buffet song. His lips touched my hair. My skin crawled as his foul-smelling breath carried the words: "Let's just get drunk and screw."

I didn't know whether to laugh or throw up. "There's not enough alcohol in the world, Tommy."

George declared the tide "high enough," and suggested that Victor try driving the boat the small distance into the cradle.

Victor put the boat in gear, and she slid ahead smoothly. Vic secured his boat to the uprights of the cradle, the only timbers above water now, with lines running from both sides at bow and stern. The Island Boys were relieved, and seemed quite proud that they had not botched the job. "We'll wait for the tide to drop enough so she's resting fully in the cradle, then we'll winch her up the beach a bit. We'll launch her again tomorrow, then we'll have time to repair your roof, Linda," George said.

"Yeah, once Vic's boat is back in the water, you'll have both the time and the lumber for the roof," I answered, not wanting the men to think they were putting something over on me. "Dad and I are going to haul traps in the morning. We'll come in early to help with the roof. See you at about this time tomorrow?" The men agreed. I waved good-bye to Victor, who was patiently waiting for the tide to drop enough so that he could wade ashore, and headed toward home.

The beer had made me lazy. I stopped beside the deserted schoolhouse. I sat and rested in one of the playground swings and dangled my feet above the oval grassless patch of ground. The wide canvas straps hanging by chains from the four-legged pipe frame had been there since I could remember. I wondered if I would ever push my own child in one of these swings. I found myself wishing I had gotten more information about Charter Boy when I'd had the chance. Did he want children? Perhaps, if things worked out, we could do our part to repopulate the Island. The visitors from the Vineyard were surely long gone by now, and I hadn't even inquired what their friend's name was. Nameless and faceless, Charter Boy was nonetheless intriguing. I wondered if other people spent as much time as I do wondering.

I'm not sure how long I sat trying to recall exactly what the woman from the Vineyard had said about her friend, whom I was certain I would never meet. I could hear the fire truck's engine now. They must be winching the boat and cradle up the beach, I thought. Charter Boy and I had something in common. Boats, islands, fishing . . . Maybe the notion wasn't so far-fetched. Suddenly the delightful fantasy I was enjoying was interrupted by a distant crash and George's voice. "Oh, fuck," I heard him shout. I remained in the swing. Soon Victor's truck came into view, and crawled up the dirt road that separated swing set and school. Vic stopped directly in front of me and spoke calmly through the open window. "Looks like the Island Boys will have to add 'fiberglass repair' to the side of their truck."

Vic smiled, inserted a stub of a cigar between his lips, and drove away leaving me amazed at his ability to remain cool. If the *Mattie Belle* had just been dropped on her side, a rock puncturing her hull at the turn of the bilge, I would be hopping mad or, at the very least, concerned. It wouldn't be until late fall when I learned that what I perceived in Victor as lack of passion was sheer composure. Victor would eventually become so livid about encroachment into our fishing zone, so tired of all the discussions, and so disgusted by our collective refusal to trigger a gear war that he would simply decide he could no longer stand to fish or even live among us.

BIPOLARIZATION

With the rudder hard to starboard, the stern scoops holes from the ocean that instantly fill back in. Circular wakes carved around blaze-orange nuclei expand outward, collapse, and settle, leaving no trace of our having been there. The lobster barrel, too, can keep a secret, as it divulges little about the effort being made to fill it in this charade that we called fishing. Like panning for gold, weeks of scooping with something quite sievelike seemed a lot of work for the few rare nuggets sifted out. It was August, and we had next to nothing to show for it. My mood had become quite black. The one thing we didn't need to worry about was the gear war. There were so few lobsters, it hardly seemed worth fighting over them.

Any normal August would find us ass

deep in lobsters. I would wake before the alarm sounded and tiptoe downstairs, careful not to disturb my parents' sleep. The first cup of coffee would taste like the best I had ever had, and I would sip it enjoying the peacefulness of a house on the water. My parents would join me before I poured a second cup, and we would admire the morning and look forward to another beautiful day. Dad and I would leave the house for a day on the water with a gourmet lunch prepared by my mother. We were three of the luckiest people in the world.

But not this year. August sucked. I hated the sound of my alarm clock, and stifled it by yanking the plug from the outlet. I stomped down the stairs, hoping to arouse everyone in the house. The coffee was bitter, and I drank it impatiently waiting for the old man to crawl out. He seemed to be getting slower and slower. I hoped we could get out of the house before my mother got up and acted happy. She just did not understand. I would rush Dad through a breakfast that he irritatingly insisted on chewing, throw some crap in a cooler for lunch, and wait for the old fart to get out of the bathroom. I would wonder, before he finally emerged, what the hell he could possibly be doing that could take so long. And I would bitch all day long.

The thought that "it is always darkest before the dawn" had me wondering how many shades of black I could descend to before a glimmer of light might sneak under the door. I was not of a mind to open the door to allow light entry. In fact, I had made a conscious decision to enjoy fully my self-inflicted depression. I would not, under any circumstances, pull myself up by the proverbial bootstraps. Hell, anyone can put on a happy face! But to truly wallow in misery and discontent takes effort, and can be fulfilling. Attempting and achieving total despair was the one thing I was doing well. In my psychological game of limbo, "How low can you go?" became a challenge.

Adjusting both internal magnifying glass and signaling mirror was key to mastering the art of lashing out at innocent victims and misdirecting anger and frustration. The most obscure remark or question could be fielded, ingested, blown up beyond life size, inspected, dissected, and reflected as gross insult. It was as easy, as they say, like falling off a log. I had, along with almost everyone else who claims to be female, fine-tuned this skill at puberty. I had also learned since then that it is best to keep the sharper reactions internal, allowing them to fester. I have been accused all my life of keeping too much inside, and been encouraged to show more emotion. To this I respond that the caring accusers and encouragers have an enduring and blind faith that there is actually something going on inside worth sharing, which is flattering but not necessarily true. I am aware that talking is a need for some. Talking is not one of my needs. I mean, it's not like food or oxygen. I would classify talking more as desire rather than need, and when I am basking in despair, seldom feel the desire.

The first civil words that I had spoken to my father in weeks were, "There hasn't been a decent dog on this island since Lucky died." He agreed, and I am sure he was relieved that my latest target had become dogs in general.

I'll always remember those first seven days of August as "the week of the dog." In my opinion, it has become very trendy to own a dog, and to possess a Lab is considered especially posh on the Island. Dogs are no longer pets. They are companions and members of the family. I've often wondered, whatever happened to "Spot" and "Rover"? Dog names now fall into two categories: people names and names that are sickeningly cute. Either way, most dog names today seem to have no relation to or correlation with the physical appearance, actions, or the emotional disposition of the dogs to which they refer. Although I grew up with dogs and cats as pets, I am not a lover of all of God's little creatures. In fact, I dislike most dogs.

Dogs can sense when you fear them, and will be more inclined to bite one who is afraid. I have heard this all of my life, and know that it is true, at least in my case. The trouble is, it is nearly impossible *not* to be afraid of dogs once you've been attacked. Many dogs have bitten me. My aunt Gracie's dog, David, likes nothing more than to bite me. David is afraid only of mousetraps. In fact, he is psychotic in his fear of them, and I suspect this psychosis stems from his having a human name. He is confused. Aunt Gracie has suggested I carry a mousetrap with me at all times to fend off David. I would prefer a baseball bat.

One of the many times David chased me from Aunt Gracie's yard barking and nipping at my backside, I sprinted home to find a message from a friend containing the sad news that his dog, Effie, had died on her way to the veterinarian. Effie had been sick with cancer for several months, and had been kept alive with doggie chemo. Effie, a Portuguese waterdog, had to be coerced with a pork chop to wade even up to her knees into salt water. She never bit me, as she had a liking for children — that is, she liked to bite them. When I first met Effie, I was not particularly fond of her, due to the fact that she was a dog. I did, however, tolerate her ass in my face when the three of us rode in the front seat of her owner's truck; I had a large crush on said owner. Effie did, over the course of some time, win me over. I was sad to hear she had died, and secretly wished it had been David.

Not long after Effie passed away, a small dog was nearly killed here on the Island when it was run over by a car. The car was driven by my mother. There is a leash law in most of the state of Maine, but the Island being the Island, we have adopted our own dog regulation. Dogs do not have to be leashed, but mustn't be allowed to run at large. An "at large" dog is one that is out of the owner's control. Two dogs came tumbling down a boardwalk into the road in front of my mother's car. She jammed the brakes, but was unable to avoid both dogs and hit one. Medical updates made their way to us. Good news, bad news, good, bad . . . My mother was extraordinarily emotional and cried off and on for three days. She was a wreck, in worse shape than the dog that had been at large. My mother was not herself, and I attributed everything to her age, and the fact that this August sucked.

About the time the injured dog was reported to be "out of the woods," and my mother had pulled herself out of her emotional spell, my older sister called to say she was coming to the Island for the weekend and bringing her dog. Her dog's name is Tie-Chee. I have always called it Chai Tea. It has since become known as Cujo. My mother expressed, quite strongly, that the dog was not welcome in her house. (During Cujo's last visit the dog had peed all over my mother's rugs.) A verbal battle ensued, ending with the slamming down of one phone or the other, not an unusual exchange between my older sister and mother.

The older sister arrived soon with the unwelcome dog and made amends with our irate mother by boarding the dog at our younger sister's place up the hill. Everything was fine for at least two hours. Then Cujo attacked my brother-in-law in his own home, biting and drawing blood. The older sister was relatively quiet until later in the day when our traumatized brother-in-law was bitten a second time, by another dog.

This time it was Dixie, a yellow Lab belonging to a friend of mine who was visiting the Island by sailboat. "She's never bitten anyone before. Except for the potato chip man," my friend explained. He was not very apologetic; in fact, he seemed suspicious of the victim. "You didn't put your hand down for her to sniff, did you?" This brightened my older sister's somewhat dismal day as she, too, could attest that Cujo had never before attacked anyone. Not even the chip guy.

The very next day, I had the honor of informing a visiting cabinetmaker that he must keep his dog, Audi, on a leash at all times. Audi had eaten Uncle Charlie's pet pheasant. Chester the clubfooted pheasant was murdered in cold blood in Charlie's yard. I was pleased. Not that the bird was dead but that the hyperactive Gordon setter would no longer be allowed to romp freely, terrorizing young children on bikes and old people with walkers.

Then came the grand finale of dog week. My father's pant leg was mistaken for a fire hydrant by the Calverts' dog, Schooner. And my father was bitten by a vicious black cur by the name of Taco. (Both in one day.) A schooner is a sailing ship. Audi is a car. Tie-Chee is an ancient Chinese exercise. Everyone knows what a taco is. Dixie is the name of a woman on a soap opera. And Effie is iffy as far as doggie names go. It was clear to me, in this bleak week in August, that dogs are confused by their ill-suited names. David? For a dog?

I was in the city of Portland last winter and witnessed a grown man, wearing suit and tie and plastic glove, pick up dog poop that was still steaming. The glove, I assume, was designed for this specific duty. It's no wonder dogs are psychotic. Diapers for babies are designed to be changed on the same schedule as oil in the car, yet intelligent people follow their dogs around and practically wipe their butts for them after they've done their business. All of this festered in my mind that August, giving my henpecked dad some relief from my previously constant: "Can't you work any faster?"

As August progressed, the lobsters continued not to budge. Islanders were now beyond panic, and we had all resigned ourselves to the dismal reality that no matter how good the fishing might get, this would not be another banner year. The best we could hope for was "fair," and that would be a miracle. The Maine lobster industry had, up until then, been an example of extraordinary success for several years, gaining attention from some unlikely sources.

Many fishermen now felt as though they had been jinxed by recent occurrences such as an article in the *New York Times* pointing out the wonders of the abundance and resilience of the lobster stock. Scientists and researchers had, for some time, been predicting quite a drastic crash in landings, and as surprised as they had been with the last annual report, knew statistically, historically, and rationally that the bonanza would surely come to an end. It did appear that the experts' prophecy was at last coming true.

I was discouraged, not so much by the poor fishing but by the fact that I had joined millions of rank-and-file Americans who hate to go to work in the morning. This loathing was new to me, and I was in a quandary about what, if anything, I should do about it. I had certainly painted myself into the clichéd corner, investing all I had by way of money, time, energy, and hope, into a life that was just not panning out the way I had dreamed. It had taken me seventeen years to outgrow swordfishing, and I enthusiastically left that world behind to return home and live happily ever after. Disillusionment took its toll in many ways.

I suddenly found that I was avoiding people. Any eye contact I allowed could result in conversation. And any conversation would probably begin and end with what I had come to call "the casual interrogation." Many of my favorite summer residents were just arriving for long-anticipated vacations and breaks from their daily grinds and frantic-paced worlds off-Island. A quick wave and "nice to see you" shouted through the open window of a moving vehicle was the no-response-needed greeting I had developed in self-preservation. Avoidance and evasion of the most dreaded questions had stunted my somewhat fragile social standing by taking me out of circulation altogether. Those bullying (in my delicate emotional state) questions, asked under the guise of genuine interest and courtesy by well-mannered seasonal visitors, tested my ability to internalize poisonous thoughts, and stifle nasty, natural reactions. But there was no avoiding Edgar Holmes this particular afternoon as he came down the dock's ramp, and I went up. Not unless I wanted to jump overboard, and I did not.

An encounter with Edgar would normally be desirable, but I stunk from bait and salt water; my hair was weirder than usual; and I was hesitant to abandon the foul mood inspired by a long day of few lobsters. "Hi, Dr. Holmes," I said pleasantly as I brushed by him. I quickened my pace as I reached the top of the ramp, and was just steps from a clean getaway when he fired the question.

"How's the fishing?"

I inhaled nervously, wondering how much detail he expected. I silently pondered my options: *It sucks. We're starving to death. What's the balance of your portfolio? Have you been sued for malpractice lately?* But instead of saying any of these, I shrugged and exhaled the one syllable "Slow," and smiled politely. I took two more steps up the dock before he took another stab.

"How are your folks?"

Old. How do you think they are? They're getting quite senile, barely know enough to come in out of the rain. They piss me off every chance they get. Hoping to appear to be in a hurry, enough to derail question three before he could ask it, I shot another syllable over my shoulder, "Fine." But the dance continued, a twisted rendition of the Mexican hat dance. I was the lowly, dejected hat, while the cheerful doctor tapped around, a free man on vacation without a care in the world.

"How's your love life?"

I then launched into a rant, savagely ripping the Island to shreds. I used adjectives to describe certain dogs that would have caused even the crudest audience to blush. I swore, and I released all my pent-up frustration and anger with the poor fishing, gritting my teeth and flailing my arms around for emphasis.

I went on and on. I gave Dr. Holmes the complete history of the Association's half-assed efforts to obtain fishing rights in our own territory, recounted our pathetic inability to launch the gear war, and told him the Island was doomed. Finally, I took a breath, and noticed that Dr. Holmes appeared to be in shock. Although I had never regarded him as overly sensitive, I thought that I may have offended him with my language or scared him with my outburst. I wondered if it might have been better to stick with single syllables. I suspected, as we parted company, that he regretted running into me. I vowed to internalize with the next inquisitor, or try to.

The next morning I was still ruminating about how better to handle the dreaded three questions as Dad and I braced ourselves for another day of working for nothing, one of the joys of being self-employed. And I began to think about other things, too — like finally getting a home of my own. It has been my lifelong dream to build a house on the Island. All I lacked was funds.

I thought about asking the Dewitt

brothers, John and Rob, to build the house for me. John's wife drives the school bus. I had a feeling that she didn't like me. She never returned a wave. The reality that the bus driver would not wave to me made me sad. So I chose to dwell on that thought as Dad and I bailed bait into totes. Why would anyone bother to build a house on this stupid island anyway? Friggin' bus driver doesn't even wave . . .

Second thoughts and self-doubt multiplied with every spiral of the all-too-familiar corkscrew pattern of hauling traps as I obsessed about everything that stood between a home and me. I began to feel like I was working on an assembly line, and could easily have been replaced by a robot. A robot would be better company for my father, who I now noticed was curiously busy behind me. I turned to see what was keeping him occupied, and was amazed to see that he was banding lobsters. The barrel, much to my surprise, was almost full, and we hadn't even stopped for lunch yet. "Well, what do you know? Payson was right. It did happen overnight. Nice . . ." I smiled genuinely for the first time all month. As quickly as the lobsters had begun, my depression fizzled.

Three hundred pounds by noontime, and I was actually giddy. I should have been embarrassed that my emotional composition could be so fickle as to do a flip-flop on cue with the lobsters. Money had never been a source of happiness before, but I was ecstatic with the results of the ongoing silent mental calculations. A few more weeks of days like this and I could pay my bills. I was indeed happy. Slightly disturbed by the realization that lobsters — ugly overgrown insectlike puny-brained crustaceans — were in control of my mood, I figured that I was in good company, for surely every other Island fisherman felt the same.

Seventy-five percent of the Island's year-round population relies on lobsters for a substantial portion of their annual incomes. Without lobsters, there would be no year-round community. I had been feeling separate, and now realized that I was part of the whole. And that whole was breathing a huge sigh of relief. Maybe even the bus driver was feeling better. My world was right again. I loved my life.

In the days that followed I began getting up early and enjoying the company of others again. I adored my parents. Hauling traps was fun. I thought my father must certainly have been convinced that I was bipolar. And then I thought, Who isn't? Who doesn't fully feel and enjoy changes and swings of mood? I supposed it was only a matter of degree that separated those who were certifiable from rank amateurs. Bipolar? Yes! And at this moment I was enjoying some time on the northern pole. My only complaint was that the days were passing too quickly.

I loved the sound of the lobsters' shells' muffled applause as they clapped against themselves and one another in the end of a trap fresh from the water. A full trap sounded like a standing ovation. I was excited by the orange and black color, like fire, as it neared the surface, of new-shelled lobsters flickering in parlor ends to be handled quickly by gloved hands, greedily filling the barrel, displacing their life's blood with mine. Fanning the fire, hauling faster and faster, the excitement crowns with the flames out of the top of the barrel and into a wooden crate that is quickly consumed. Fishermen feel a frenzied need to fuel the fire; once extinguished, history has proved it might not reignite until next season or perhaps ever. All of the work of kindling, fanning, stoking, and heartbreaking false starts had finally caught. I was, along with others, manic. It seemed that everything in proximity of the fire was brightened by the light cast and warmed by the heat produced. I began to think that I was where I wanted and needed to be. The Island would always be home.

Emotion, like water, seeks its own level. Without being dammed up or forced in one direction by a pump, balance is found. I settled comfortably into the Northern Hemisphere of my bipolar being, just above the equator, and I liked the climate there. Lobstering was consistently good, not the best, just good. And good was great. The weather was good, not flat calm and clear every day, but good. Life was good, not perfect, but good. It could go on forever, but, of course, it would not. Something would upset the balance eventually, and I would have to start over. This grasp on reality was what kept me from joining the truly psychotic, I thought. The days were getting shorter. The lobsters would soon move to deeper water and disappear. The weather would sour. The fire would go out.

But the weather stayed good. The fishing stayed good. Life stayed good. Before I knew it, weeks had passed. Dad and I cleaned up the *Mattie Belle* especially well, left her on the mooring, and headed home a bit early one particularly fine day to find out what goodness my mother might have on the stove. I barged through the door ahead of my father as usual and yelled, "Hi, Mom! We're home!" I was disappointed not to smell something delicious as I had anticipated, perhaps even taken for granted. Mom was in the kitchen, but she was not cooking. She looked tired. Or I thought, maybe she had been crying again, not an unusual occurrence as of late. Being my usual smart-assed self, I asked, "What? Did you run over another dog?"

"No. I have breast cancer."

LENGTHENING OUT

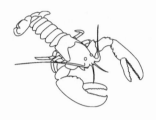

" 'Bout given up in the shoal water, ain't they?"

"Ain't it awful?"

"I'm off on thirties now. Not much to it. I'll lengthen out a few more and hope for a spurt."

"Yup. Put the warps to 'em. That's all you can do. That, and pray."

As the days got shorter, the warps got longer and the lobsters fewer. Hearing this conversation on the VHF radio as we hauled our traps, I was perplexed about what to do. All my gear was in 10 fathoms of water, while most everyone else had moved theirs offshore. Conventional wisdom says that lobsters head to deeper water as it gets colder, so you are supposed to shift your traps out accordingly. I probably should have shifted out to 20 and 30 fathoms with the others, but I had not, and now the reports of lack of lobsters in deeper water discouraged me from the work of lengthening out at all.

Dad and I were content with what the traps were producing this morning, nearly 2 pounds each. But I reminded myself we had not hauled a single trap in over two weeks. Neither of us had the heart to work after we learned of my mother's diagnosis. On the day Mom told us she had cancer, everything in my view was suddenly diminished, as if I was suddenly looking through the wrong end of a pair of binoculars. I couldn't give a damn about lobsters or anything else while we waited to learn more about the lump in my mother's right breast. The two weeks had been both emotional and weird. My gear had not been lengthened out, but my perspective had.

The long soak had certainly filled the traps with lobsters. We were filling the barrel quickly, but I had a sinking feeling that if the traps remained here along the shore, we would catch absolutely nothing when we hauled the next time. On the other hand, I rationalized, it didn't sound as if lengthening out was paying off for the other fishermen, so I might be just as well off fishing my traps right where they sat until "taking them up" for the winter. Reports up and down the coast all confirmed that landings were down 30 percent most everywhere. I didn't much care.

My mother had left the Island for a lumpectomy and was now home waiting to hear from her oncologist as to what the follow-up treatment should be. My mother had urged Dad and me to resume our daily routines. We tried. The monotonous routine of setting and hauling was somewhat reassuring. "Mom has breast cancer. Look at us, catching lobsters just like we didn't know any better."

"This is much better than I thought it would be," my father replied, pulling a keeper from the main trap. I was not sure what Dad referred to as "better" than he had anticipated: the fishing or the fact that we were, for the first time in fourteen days, doing something in addition to waiting, wondering, and worrying.

"We'll have to lengthen some out tomorrow," I said, making the decision to do it just as the words formed. It was what my mother would want. As I recalled seasons past, it seemed a bit early for the lobsters to be heading for deeper water. Maybe that's why the other fishermen hadn't yet had much success farther offshore. But there was a hint of fall in the air, especially in the mornings. Other indicators of autumn's arrival were bolder and harder to ignore. What I had noticed first were the deer.

The bucks, who for most of the year are far less sociable and visible than the females, had been out and about for two weeks now, a month earlier than their normal rut. The Island takes on a strange and remarkable beauty in the weeks between summer and fall, when the foliage remains green and the cranberries that dot the bogs turn a deep crimson. It's my favorite time of year. The week before, when my folks had been off-Island for Mom's appointments, consultations, and tests, I had decided to go for a long walk and had an experience that was now haunting me.

At the risk of sounding like a baby, I had been missing my parents. I didn't like the feeling of living in their house without them, so I had been spending as much time as I could out of the house. I had heard that a dead whale had washed ashore on the beach above Eben's Head, and decided I would take a hike and see it myself. I had in mind that I wanted to know if the smell was as bad as the school kids had reported. (No museums or art galleries on the Island meant school field trips to see a dead mammal on a beach.) I did not know why I had a desire to confirm their takes of the stench. "It was gross." "I almost threw up." "It made my eyes water." But I did.

In four hours of hiking and beach-combing, I failed to stumble upon the decomposing pilot whale. I kept sniffing the air but smelled nothing offensive. I figured the last high tide must have floated the carcass back offshore. Unwilling to return to the empty house after the unsuccessful mission, I decided to continue to hike and enjoy the day. I walked back to the main road, where I had left my truck, jumped in, and drove to the west side of the Island, through town, and to the trailhead marked "Mt. Champlain." I had now more than circumnavigated the Island and had not seen a single human being. I had not passed a vehicle on the road or greeted a hiker on a trail or waved to picnickers on a beach. It seemed even more desolate than off-season last year when I saw the same six people every day for months, and thought I knew how Gilligan must have felt. Now it seemed I had the entire island to myself.

I had not climbed to the top of Mt. Champlain in twenty years, I thought, as I tied the sleeves of my sweatshirt around my waist.

The first third of the gradual ascent was by a well-marked path through a relatively thinly treed forest of foreign-looking pines. Ninety percent of the Island is host to families of Sarah Orne Jewett's "pointed firs," the triangular-shaped spruce trees we decorate for Christmas, with short needles and branches that are longest at the base of the trunk. Here, at the Island's interior and higher ground, grew long-needled pines with umbrellalike forms that the trail leads hikers under rather than through.

It took me about thirty minutes to reach the summit of ledge and bay leaves that were well on their way to peak foliage. I was on top of the Island, the pinnacle of my world. To the south stretched nothing but acres of Island. Although the trees had grown higher since my last visit, somewhat obstructing the view, I could see water to the east, Camden Hills across the bay to the west, and to the north, beyond water and many smaller islands, the town of Stonington, 6 miles away. All around I could see hardwoods poking colored holes through the blanket of green. I never knew there existed so many shades of red. The sole witness to all of this beauty, I felt lucky, even blessed.

As I began my descent, I thought of my mother, and how much she loves the Island.

Then I imagined my mother bald, and knew she might be the next time I saw her; biopsies of lymph nodes had revealed bad news that required aggressive chemotherapy. I wondered if she would look at all like the infant I held in my mind's eye. She would be mad, I was sure, if she knew how much time I was spending worrying about her and imagining her bald.

My mother would need to go back and forth to the mainland for treatment, spending weeks at a time off-Island. I was sad, realizing that she would miss most of her favorite season. The ferns, now golden, would be brown by the time her chemo was done. The reds would all have turned to scales of dry rust and fallen to the ground, where the next rain would coax them to disintegrate into the ground and take on a mushroom smell that would be freshened with the first snow.

I picked the most vibrant leaf from a bush beside the path to send to my mother. Then I saw another I had to get, and another, and another, until I wandered with a handful looking for yet another more stunning than the last. I had strayed far from the trail. I stopped. Which way back? I heard only my own breathing.

I experienced what I guess was an anxiety

attack. My heart fluttered in my chest, and there was a slight throbbing in my throat. Although I had been perspiring from hiking, I felt a chill. I became panic-stricken. Oh, not that I was lost. After all, it is an island. How lost could I get? I knew that if I walked downhill I would eventually find the main road, and there was not an inch of its 13 miles that I would not recognize. But I was suddenly frightened of being alone. What if I fell and broke a leg? Or stumbled into an old well or mineshaft, like in a movie? Who would know? It could be days before anyone got suspicious. Would anyone search for me? Would I be missed? By whom? I felt like a forty-year-old orphan. I might perish right here. My decomposing body might be discovered by Greg Runge's dogs. Maybe the teacher would arrange a class trip . . .

"Jesus Christ! I am absolutely foolish," I said aloud (I know it's an odd thing to say) as I threw the leaf collection to the ground and started downhill. (The leaves would turn brown, I remembered, even if I shipped them in a plastic Ziploc.) I also remembered that Aunt Sally would certainly note my absence at her dinner table if I did not appear. She would send Uncle Charlie to find me. Charlie would see my truck at the bottom of the trail and know where to look for me.

There are no mineshafts on the Island. So I realized that this was what lonely felt like. The anxiety passed as quickly as it had come. But it crept upon me from time to time for weeks after — even as Dad and I continued to haul and pretend that all was fine, neither good nor bad.

Most people held that a drought we were experiencing was responsible for the early arrival of fall, which made sense with regards to the foliage. But would the lobsters pay attention to the drought, I wondered? How could a dry spell affect them? It made no sense to me, but the next morning Dad and I gathered coils of 20- and 30-fathom pieces of floating line with which to lengthen out. We would, indeed, per matriarchal orders, resume daily routines. We would not let on to the lobsters that anything was wrong.

Twenty starboard circles into our day, we had hauled, picked, baited, and stacked onto the *Mattie Belle* the first forty traps to be taken for a short boat ride to deeper water. I steamed west, off the west shore of the Island until the depth sounder marked 25 fathoms, or 150 feet. Now it was painfully obvious that I was far behind the crowd in shifting to deeper water. There were buoys everywhere marking traps others had set. It was also obvious that my father and I were paying strict attention to detail, and all conversation was of lobster and traps. There was no discussion of my mother's health. We both seemed to prefer it that way.

As I poked around for a hole into which to set a few, my father lengthened out the pair of traps that sat on the gunwale. He untied one of the 20-fathom coils that we had brought aboard that morning and cleared both ends. Next, Dad untied the double sheet's bend securing floating line to pot warp, and inserted the lengthening piece by tying each of its bitter ends to the ends produced by untying the original knot. I searched for some room between buoys to set just one pair. Dad waited for a sign from me to push the first trailer overboard. I zigged and zagged my way south, but was frustrated by the lack of room.

I jogged the boat to the west again, heading for water a bit deeper. The pair of traps that waited patiently with my father at the rail were rigged to be set in water as deep as 180 fathoms. When I reached the maximum depth of water for the lengthening pieces I had on board, I again ran south in that depth looking for a spot clear of others' gear. Finally, I saw wide-open ocean ahead of me and went to the clearing. Odd, I thought, that there was no gear in this area at all, as far as the eye could see. Suspicious, I checked the plotter and even pulled out an old paper chart to ensure that it was legal to set here. Confident that I would not be arrested, I nodded to my father to set the first pair. I was, I knew, pushing my luck as far as unwritten rules and boundaries observed, but what the hell. No one else was fishing here. Why should I worry about a few lobster traps?

The bulk of the Association's membership was not concerned about anything, it seemed. They didn't want a gear war and they didn't want to try to pass legislation to protect our turf. They just chickened out, I thought. Ultimately, it was fear of reaction or retaliation by those who would have been excluded from our zone that caused the Association to give up. Jack's health was failing, and I doubted that he had enough steam left in him to rally the troops once again. And what was the use? Every time we decided something, we undecided it weeks later. I had been right, as it turned out, in hesitating to join the Association. I was not good at group membership and vowed not to get caught up in it again.

Along we went, slowly to the south and

east, until the deck was clear and the last two of the forty traps waited for the nod. "Okay, Dad" was followed by a splash, 10 fathoms of yellow floating line, and the second splash caused by the trailer trap. I watched the coil slowly unwind from the deck and over the rail as if some magical snake charmer seduced the line from its relaxed loops. The buoy then plopped onto the surface, rocked as the wake of the boat passed under it, and finally came to rest, steadied by the weight of the warp.

We had traveled so far to find a spot to set the longer warps that daylight allowed only two boatloads of gear to be shifted. It is illegal in the state of Maine to haul traps after official sunset. When the eightieth trap hit the water, we were content to have accomplished a day's work and were anxious to get home to my mother. She would be glad that we were acting normal. In fact, I thought, Dad and I could be nominated for Academy Awards.

Darkness had gained ground by the time we cruised between the two outcroppings of ledge that mark the narrow entrance to the thoroughfare. I idled the *Mattie Belle* toward the mooring; it was now quiet enough to hear yourself think, so I thought out loud about the area where we had just lengthened out. "It can't be good. If there were a friggin' lobster there, the place would have been buried with buoys. I don't expect much."

"Put the warps to 'em. That's all you can do," Dad responded, imitating the voice we had heard on the radio the day before.

"That, and pray," I added.

THE FOGHORN

On a brutally cold, late morning, mid-January in the early 1900s, Mattie Robinson, my father's mother, stood in the kitchen with her mother, Lillian, opening mason jars of blueberries that had been "put up" in August. The berries, almost black in color, glistened in their own syrup like overgrown caviar. Mattie, who had not yet celebrated her thirteenth birthday, recalled what her mother had said in defense of all the work involved in picking and canning berries in those last summer days when the girl would have much preferred rowing and drifting wantonly around the cove in her cherished boat. Lillian had declared, *and now it was true,* that on the coldest day of the winter, *which this certainly seemed,* the women of the house, *which included mother and this only daughter,*

would lovingly prepare steamed blueberry pudding and a cobbler that would warm both body and soul.

The wind was northerly and individual gusts shrilled as they bent around the corner of the only house in the cove. A small puff found its way down the stovepipe, fanning the fire and adding a crackling accompaniment to the dull and constant howling. Mattie watched through the window over the sink, the vapor thick and low on the water, rolling like smoke from the chimney, from the right to the left border framing her view. "It even *sounds* cold," Mattie shivered. The snap of the metal fastener against the glass, followed by the popping of the lid's rubber seal from the jar, drew the young girl's attention back into the warmth of the kitchen.

As Lillian dumped the jar's contents into a large mixing bowl with the other ingredients, she completed a thought aloud: "like opening jars of summertime." The wooden spoon swirled summer from a lumpy ink to a lush field of purple lupine. Mattie hummed while she stirred, and her mother greased and floured a baking pan, shaking and tapping the pan against a palm to coat every corner. Lillian suddenly held the pan very still and cocked her head to one side, listening. "Did you hear that?" She asked.

Mattie stopped smoothing the batter and perked up her ears, straining to hear something other than the wind. There it was: not a sharp knocking at the door but more of a muted thumping, like the repeated striking of a dead piano key. "Who would be out today?" Lillian wondered as she wiped her powdered hands on her apron. Mattie abandoned the spoon, half submerged in the batter, and followed her mother to the door. Mother and daughter were both startled by the sight of what began on the doorstep and continued into the snowy yard: A group of men, some supporting others who could not stand alone, stood and looked hopeful through thirteen pairs of exhausted eyes. The man in front, who had been knocking with a fist balled up inside a frozen wool sleeve, spoke through an icy beard. His words were urgent, and his tone convincingly plaintive. Decades later, all that Mattie remembered of the explanation that got the men on the warm side of the door, was "shipwrecked."

The captain and the twelve-man crew of the ill-fated fishing schooner had sailed from Gloucester Harbor the previous day, enjoying a brisk westerly wind. Bound for fishing grounds south of Newfoundland, the men were full of excitement for, and anticipation of, hooks heavy with cod, the tubtrawlers' bread and butter. The younger members of the crew spoke greedily about how they would spend the pay they already imagined in their pockets. The weather turned sour very quickly.

Soon they were in the midst of a howling northeaster, and a blinding snow squall. It was then that the captain decided, for the safety of his crew and vessel, which were both being wracked by the storm, to try to find safe harbor, a lee from the seas that threatened to pound men and boat to pieces. The southwestern and leeward shore of this mountainous island would have been the ideal place to anchor and wait out the gale, if it hadn't been for the ledges that peppered the area. From Western Ear to Trial Point, vicious ledges lay just beneath the surface, while others boldly poke their heads above. These remote outcroppings of rocky peaks are surrounded by deceivingly deep water; some rocks are as far as a mile from the coast. The men, convinced that they were doomed if they remained at sea, took their chances at navigating the treacherous gauntlet.

The initial strike wasn't that bad. A single sharp jab resulted in one cracked plank, then the boat was again afloat and under way. Minutes later, a second crash onto a ledge that was not as merciful caused the men to question the decision to seek relief from the weather. The schooner rose, suspended for a silent second or two, then crashed down again and again, each surge severely damaging the hull that was now steadily taking on water. The boat was bound to break up. Daylight crept over the highest point of the island that loomed so close yet was a deathly distance to swim in such frigid conditions.

Something changed at daylight — wind, or tide, or both. The sinking vessel listed to starboard, freeing her from the ledge, and she wallowed slowly, lying mostly on her battered side, toward shore. Through a lashing freezing spray, it was now light enough to see the shore outlined in frothy foam. It became abundantly clear to the captain that there was no saving the ship; he could scarcely find any hope or reason to believe that he and his crew would not perish. They would surely drown, be beaten to death on the rocks, or freeze solid while waiting for the drowning and beating.

It was clearly a miracle when the captain counted thirteen heads, including his own, staggering above the high-water mark and crouching in the first stand of trees. Somehow all had made the swim to shore. So far, these men had cheated death. But battered by the rocks, they were in very bad shape.

The conclusion of a short debate on how to proceed was that they must stay together. Anyone who stayed behind while others, stronger, went for help would certainly die if assistance was not around the next bold headland of sheer cliff that lay north of the group — the direction toward which they all agreed to trudge.

Some walked, some limped, two were nearly dragged along cobbled beach, over ledge, and through woods where the cliffs were too steep to pass. Progress was pathetically slow. After one discouraged soul pointed out the short distance they had traveled, the captain barked an order not to look back. The storm continued. The men were soaked. The temperature was freezing. The men were stumbling along a deer path, through a thick forest of spruce, when one of them claimed to have gotten a whiff of wood smoke.

The fishermen staggered toward the smoke, ducking under branches heavy with snow, and finally broke out of the woods and back onto the beach in Robinson's Cove.

Here they had full view of the house with the chimney, in which they were soon to take refuge. Shortly after they arrived, so did Mattie's father, Charles Robinson, who had been off feeding the sheep he kept in a barn on the hill.

The men were too weak to assess their own physical damages when Charles Robinson left the house early the next morning. As soon as the daily mailboat arrived, he would send word ashore of the shipwreck and survivors. The captain of the mailboat would bring the news to Stonington, and it would soon reach the fishermen's home port of Gloucester, Massachusetts. Another boat would be sent to carry the thirteen home. In the meantime, ten days, the men would sleep and eat, the only two remedies prescribed by a doctorless community for frostbite, hypothermia, and the numerous wounds that had been caused by the rocks. "Fish tea," a hot stew made with salt cod and potatoes, was Lillian's version of chicken soup, and a cauldron of it simmered on the back of the cast-iron stove through the entire convalescence.

Toward the end of their stay, on a fair day, two of the Gloucestermen returned to the shore where they had arrived and retrieved the only physical evidence of their having been washed up there. The schooner's fog-horn, a wooden box sporting a brass lever on one side, sat under a tree as if placed there by the hand of God. There was no flotsam swishing back and forth in the tide, no rigging strewn along the beach, nothing else.

As a gesture of thanks, the men left the horn when they departed. My great-grandmother soon took to using the fog-horn to call my grandmother Mattie home when she wandered late. My grandmother, when she became an adult and had children of her own, adopted the practice and would crank the lever to call my father and his four siblings home in the evening. And my sib-lings and I had all pulled the lever back and forth through the years to produce the loud bellowing sound that had withstood the wreck and, more amazingly, three genera-tions of curious children. The leather strap handle still bears the salt stains of eight de-cades.

Now I sat with my feet propped up on the foghorn and watched the vapor, thick and low on the water, roll like chimney smoke from the right to the left border framing that same view. The Island really had not changed much since my grandmother was a girl. My grandmother loved to fish. My mother does, too. And if there is one point of knowledge common to all fishermen, it is this: Things can quickly go wrong at sea. The foghorn, however, is always there to remind my family that shipwrecks (as well as other calamities such as a terrifying medical diagnosis) are not always fatal.

THE LITTLE LOBSTERMAN

Nicholas Robinson Barter is all Islander. His bloodline is more thoroughbred and undiluted than any other claiming such status. Nicholas' family tree is rooted firmly in granite, spruce, and salt on both maternal and paternal sides. A direct descendant of Peletiah Barter, the Island's first permanent settler, Nicholas is seventh generation. On his mother's side, he is heir to Robinson, Hamilton, and Bowen, who were not far behind Barter in being granted chunks of Island and making themselves at home. Prior to the first settlers, offshore islands had been seasonal hunting and fishing camps for Indians. The French and British who came to the New World started the first commercial industry offshore, using islands as remote fishing stations. In 1792, just after the Revolutionary War, Nicholas' ancestors became the Island's first year-round residents.

At the age of nine, Nicholas has no intention of forsaking his three-dimensional congenital heritage, and is the hope for the community's future. Over the years, as children on the Island have come of age, it has usually been the case that the boys settle on the Island and become lobstermen and the girls leave to marry. In keeping with Island tradition, Nicholas fished his first lobster traps (five of them) at the age of six. When he was that age, his mother had the most difficult time tearing her son from the water's edge. Nicholas was constantly slogging around in wet sneakers; even when he wore boots, he managed to wade in over the tops.

Until Nicholas was nine, no one ever wondered where he was because he could always be found in one of three places: knee deep in water at Collin's Beach; jumping in and out of skiffs at the dock; or hauling his traps from a skiff in the thoroughfare in the company of any willing adult. Nicholas rarely missed meeting an arriving mailboat and became a miniature harbormaster, graciously greeting and directing visiting vessels to rental moorings.

At six or seven, Nicholas had become the youngest member of the Lobstermen's Association. He attended meetings with a notebook and pencil. Nicholas is genuinely a nice boy. Everyone is fond of him. So his presence at Association meetings became a reminder of the need to secure fishing grounds for the survival of future generations. Nicholas quickly and quite naturally became the reason for all members to at least talk a good game about gear wars and legislative battles.

But by now Nicholas has lost his baby fat and his interest in the water. He is a very kind and happy kid. He skips up and down the road and sings to himself. He likes Pokémon and Game Boy. He is proficient with computers and is a good student. He is an avid reader, most recently of the Harry Potter books, and experiments with a chemistry set. I asked him recently what he would like from me for an approaching birthday. He responded, "Oh, I'm all set. Thanks." I couldn't believe my ears, and prodded him for a list of wants and needs. He was steadfast. He wanted nothing.

Nicholas has not hauled his lobster traps for over a year. (His buoys, undisturbed except for when joggled by boat wakes, had so much growth attached to them they were barely buoyant.) With all his interest in books and computers, he must have grand ambitions, I figured. The Internet has opened new realms of possibility to Islanders. Nicholas has a keen knowledge of astronomy and natural history. So I asked, "What do you want to be when you grow up?"

He did not hesitate. "A lobsterman, just like I am now."

PERSEVERANCE

Nicholas Barter's future as a lobster fisherman was unquestionable, at least in his nine-year-old mind. At the age of forty, I was lukewarm on lobstering and growing increasingly unsure about my future in it. I was certain only that I wished to remain on the Island, as it was the only place other than the ocean that I found comfortable enough to regard as home. I wondered what I could do to support myself if the lobsters suddenly became so scarce that fishing for them was unprofitable.

I also worried that the Island, due mostly to everyone's uncertainty about the future stock of lobster, might begin to change at a quicker pace. If the Island should someday be inhabited only by caretakers of summer folks and summer folks themselves, the Island would surely suffer an identity crisis. Who would want to live here then? The two populations, summer and year-round, now enjoy a symbiotic relationship unlike any of which I have ever been aware. We do not refer to seasonal residents as "summer complaints," nor do they have derogatory names for us. We are all Islanders when on the Island. Some of our summer population had been coming to the Island for generations before many permanent residents discovered life here. Although divided by a financial canyon, both populations are united in wanting the Island to remain unchanged. We all want to thrive on our resource-based economy and not cave in to tourism. Islanders love the Island just as it is.

My siblings can all be considered seasonal residents, as they come and go mostly in the months not involving snow. They love the Island as much as I do, but they have real jobs — careers, professions — and lives that require bases on the mainland. I do not possess the tenacity, drive, and education that allows my younger sister, Bif, to excel in corporate America. Nor do I have the education, licenses, knowledge, experience, or skills of my brother Charlie, an engineer who designs and installs commercial-grade boiler systems. And I could certainly never be versatile and well-rounded enough to gain membership in "the job of the month club" like my very adaptable and multi-talented sister Rhonda. I do not want to be my siblings any more than they want to be me.

As different as my siblings and I are, we also have a great deal in common. And one of those things is that we all — in our own ways — pray. Following my mother's diagnosis, the focus of my prayers shifted from thanks *to* to requests *for*. I was disappointed that my requests were not being fulfilled, and was most uncomfortable with the role of supplicant. I did not like asking anyone for favors. The act of praising and thanking for bountiful fishing, beautiful weather, and safe travels at sea had always been uplifting. Now my nightly solicitations of my God requesting improvement in my mother's health and more lobsters left me with a feeling of weakness.

Prayers were not working, I thought, as I broke two empty traps over the rail. Perhaps I needed more patience with God. The request line was surely much longer than the worship and praise line. Everyone wanted or needed something, as when, at the deli section of a crowded supermarket, I took a number and waited. I could control neither lobsters nor cancer but believed someone or something could and did. So I tried every night to get to the head of the line.

I hauled traps alone on this day, as I had also grown impatient with waiting for my father to return to the Island. Overnight our business arrangement had slipped from a partnership to a sole proprietorship. Dad and I had an agreement, more like a pact actually, that neither of us would ever haul traps alone. Although other fishermen worked solo, it was widely understood that the practice was dangerous. Ben MacDonald's father and the man across the bay who was found wrapped in his own propeller gave evidence of that. Hauling alone was not only a broken promise to my father but also foolhardy.

My father had on several occasions taken my boat and hired a temporary sternman to haul traps when I was unable, unwilling, or unavailable. But I chose to go it alone. This breach of agreement would have infuriated me had the roles been reversed, yet I justified it many times over. The traps had not been hauled in over three weeks. There was no bait left in the bags, and the trapped lobsters were surely cannibalizing one another, which was wasteful. Every able body on the Island was otherwise employed. The water was flat calm. I would be extra careful. I needed time alone to think.

What I thought most about while I worked was *why* my prayers were being ignored. Hell, I had read numerous accounts of miracles granted. And what I wanted was certainly short of miraculous. All I asked was for my life to return to normal. I wanted my God to intervene with a bolt of lightning or something equally dramatic. I must be doing something wrong, I thought, as I continued hauling, picking, baiting, banding, and setting.

I wondered how other people prayed. It was far too personal to ask, even of one of my siblings. If I could ask anyone, I supposed I would ask Payson Barter, who had just steamed by me with a load of traps. My own nightly prayers, which I suspected were somewhat unorthodox, took the form of a letter dictated by me. The letter always opened with "Dear God," and closed with "Sincerely yours." I prayed lying on my back, with my eyes wide open and arms folded across my chest. Perhaps, I thought, I should close my eyes or bow my head. Maybe "Sincerely yours" was too formal. I was unsure. I tried to think of a way to eavesdrop on Payson's prayers.

Payson is the Island's top lobster producer. The key to his success is common knowledge and is printed across the stern of his new boat for all to see: *Perseverance.* I knew it as well as my own name. I had spent my life learning it. Hard work, persistence, and determination had taken me everywhere I had ever been. As the stern of Payson's boat faded in the distance, I thought back to my junior year at Mt. Ararat. Near the top of my class academically, I overheard my favorite teacher refer to me as "the classic overachiever." I was deeply hurt and insulted by the reference. I had, at seventeen, always considered myself rather brilliant. "Why," I thought, "this teacher thinks I'm just a big dummy who works hard. I'll show him!" Well, the past twenty years or so had proven the teacher quite perceptive. I am not smarter, stronger, or faster than anyone I know. My one asset is my ability to work, and I have never been afraid to exploit that.

If I had paid heed to the well-meaning advice and suggestions from loved ones, I would have "gotten a real job" years ago. I had not missed a cue — of that I was sure. I was not intended for a real job, not everyone is. I remembered again my mother's initial reaction to learning that I was taking my Colby College diploma to sea aboard a swordfishing vessel. As bad as it had been, I would have given anything to see her smash a few stacks of dishes now. Then I would know she was fine.

Watching her fight against cancer, I learned how apt the word *fight* is in describing what cancer patients go through. I had great confidence in my mother, but I had never imagined such a brutal boxing match. The first round with chemotherapy was one in which my mother had not fared well. Waves of nausea resulted in a standing-eight count. Dazed and confused, my mother wobbled around the ring unsure of whether to swing back or fall to the mat. Suddenly she went down hard in mid-round, landing in the hospital for four days. The poison had dealt a blow that took her off her feet. The oncologist called it "neutropenia" and explained that the bone marrow was not producing white blood cells fast enough to replace what the chemo was killing. I was told that the normal white cell count was eight thousand and that my mother's went down to four hundred during the middle of her treatment cycle.

It was such a helpless feeling to watch my mother get pummeled and know that all I could do was pray for the bell to sound the end of that round. When the bell did ring, my mother was able to limp to her corner and regroup for the next dose. Her blood count recovered, she felt good, and she was bald. As distinct from Goliath, the loss of hair gave my mother renewed strength. One very bald mother shook a finger at her daughter and instructed her to go home and go about her business. The daughter obeyed, for this was the only time her mother had ever ordered her to go fishing.

I laughed out loud. I thought that if I could collect a dollar bill for every time over the past twenty years that I had been told that I was "wasting my education" by going fishing, I would have far more money than the lobster barrel would produce by the end of this day. Which some would say was evidence against my assertion that education is never wasted, but I had never measured my success in dollars and cents. Financial status is no measure of my achievement, any more than SAT scores and class rank were indications of what kind of student I was and still am. I am a product of my education, both formal and incidental. I have used every bit of my education every single day, fishing or not.

Because of my formal schooling, I knew things like one plus one equals two, Abraham Lincoln was the sixteenth president of the

United States, and *i* before *e* except after *c*. Incidental education had taught me to recognize the value of persistence, determination, teamwork, leadership, confidence, loyalty, honesty, trust, responsibility, accountability, and faith. . . . I cringe at the phrase "education is a tool." I don't think it is. Tools are designed for very specific purposes, like a gaff, for instance. When not employed hooking buoys, the gaff lays dormant, wasted, like mine, which now lay on the rail. Education is always being used.

By mid-morning I was worn out from the frenzied pace of trying to do the work of both skipper and sternman, and had barely enough lobster to cover the bait and fuel expense. While fishing alone was indeed good mental and physical therapy, I could see that a sternman more than paid for himself with the increased production. The going rate for a sternman is what I had paid Stern-Fabio, 20 percent of the gross stock. With the help of a good sternman, you could double your catch. It was a no-brainer. Gazing into the barrel, I estimated 30 pounds of lobster. "Holy shit," I said as I tried to figure the number of traps I had hauled.

I had used just less than one tote of bait, so I guessed I had hauled around sixty traps. Half a pound per trap was about as bad as it

290

gets, so I decided to steam to another area to try my luck. The traps I had handled so far this morning had been in shoal water and rocky bottom. I headed for deeper water and muddy bottom, and prayed for more lobster.

As I steamed the *Mattie Belle* to the southwest, I lamented the fact that this season had fizzled the way that it had. I supposed that the fishermen who had been diligent and had not missed any possible "haul days" had fared well. Payson, for example, had probably done fine in spite of the late arrival of shedders. I had not bothered to tally up my lobster receipts for the season, and knew that when I did I would be ashamed of myself. I had certainly missed many opportunities. "Damn cancer." No sense comparing myself to Payson, I thought; Payson is in a league of his own. I would be wiser to compare myself to George and Tommy of Island Boy Repairs.

George and Tommy had set a new standard for poor fishermen the year they decided to give lobstering a shot. They did so badly that their repair business was in comparison on the Fortune 500 list. Even the school kids hauling out of skiffs caught more than the Island Boys. Neither of the men had lobstered prior to the announcement that they would become lobstermen, and I am sure that I was not the only one wondering how they finagled licenses. They had been adequately warned about buying used traps but managed not only to ignore the warnings but also to purchase the junkiest pile of antiques any of us had seen. They committed cardinal sin number one, using wooden traps, the kind that had been employed only as coffee tables for the past two decades. They rationalized, "Well, these traps *used* to catch lobster." Funny, we all thought, that they bought gear before they bought a boat.

When their boat appeared on a mooring, everyone agreed that it was as bad as the traps. It was no wonder they failed to make any money. They spent all of their time and resources repairing boat and gear. On the odd day that the boat's engine did start, they had to be towed back into the thoroughfare before they were finished hauling. And towing the boat was tricky. Old wooden boats that have not been maintained tend to rot. Rotten boats are referred to as "tender." She was certainly tender, and as a result of towing needed additional work by the Boys. George and Tommy learned the hard way that equipment is an important ingredient in any successful fishing operation. The biggest part of the Boys' education had been incidental, I thought.

Payson had learned the lobster business from his father. I had never seen Payson doing anything but work, and knew that he had learned his strong work ethic from his father. Payson never appeared to get too excited or bent out of shape about fishing, so it was hard to tell if he enjoyed his work. But he surely must have liked being the Island's top producer year after year. As much as I like to think that the volume of lobster caught is a function only of the number of traps hauled, Payson proves it isn't. The *Perseverance* was rarely the first boat out in the morning or the last one back in the evening, yet Payson invariably had more lobster than anyone else. As I steered the *Mattie Belle* toward a blaze-orange buoy, I again saw the *Perseverance* in the distance, keeping her usual slow and steady pace. Payson never dashed around making noise and commotion. I gaffed the buoy and hoped for lobsters.

The Island Lobster Association had been "pounding" lobsters since the first of September. Every season after Labor Day the price of lobster drops significantly due to changes in demand. In the fall, Vacationland becomes less so. Fewer visitors to

Maine means less demand for lobster. The supply, however, is steady during September and October; so, of course, the price dives accordingly. The price recovers prior to Christmas due to huge holiday demand and dwindling supply. Generally the supply reaches its lowest point in the months of February and March, and those dates see lobster at the highest prices. Fishermen who brave the winter elements are rewarded by receiving prices often in excess of $6 per pound, as opposed to the $2.75 that I anticipated this day in October.

Pounding lobsters is the act of storing lobsters in a "lobster pound" during the months when the price is low; you can then sell them for greater profit in late winter or early spring. On the Island, we are fortunate to have the use of a natural pound, a saltwater pond in which we have at times stored up to 40,000 pounds of lobster while waiting for the price to go up. The pound is equipped with an electric pump that aerates the water, protecting the lobsters from suffocation. There is some tidal flow in and out of the pound, keeping the water from stagnating. Pounded lobsters are fed bait and medicated feed to keep them healthy and free of disease while in storage. Once the price is declared "as high as it will get," which is guesswork by the Association's management, the work of taking lobsters out of the pound begins.

Some of the members of the Association are certified scuba divers. I'm not one of them, thank God. The divers don dry suits, masks, fins, and weight belts and plunge into the freezing cold pound to retrieve stored lobsters. The divers swim along the bottom, grabbing lobsters and placing them in mesh bags. Once a bag is full, the diver marks it by releasing a small float that is attached to the bag and pops to the surface. Two men in a skiff haul full bags aboard and transport them to a float where other fishermen empty the bags into crates, weighing them on a scale to 90-pound units. The 90-pound crates are loaded onto a lobster boat and taken to Stonington, where they are sold to lobster wholesalers, who in turn sell and distribute to retail outlets worldwide. Somewhere in the midst of storage and changing hands from producer to consumer the value of a single lobster may go from three bucks to over twenty.

Although what lobster I caught today would go into the pound to be sold six months down the road, the Association would pay me the going rate of $2.75 immediately. If I chose to, I could sell my catch on the mainland for an additional fifty cents per pound, but by doing so I would forfeit my "bonus." The bonus would be my share of the money divvied up among Association members when profits are realized in early spring from pounded lobsters. In the last five years the smallest annual bonus paid to members by the Association was sixty-five cents, a net gain of fifteen cents for every pound caught and sold all season. The timing of the bonus makes this "forced savings" particularly appealing, as bonus checks are received in April when all fishermen are settling up with Uncle Sam and regrouping for another year.

As I approached the "lobster car," a raftlike float where I would leave my catch for the Association managers to weigh up that night, the *Perseverance* crossed my bow ahead of me bound for the mooring. This was the fourth time I had waved to Payson since this morning. I placed about 70 pounds of lobster in a crate, tied my buoy to it, and pushed it into the water where it would hang tied to the raft until later that evening. Whatever bonus I received next spring would be needed, I thought, as I washed down the deck with the hose from the barrel. I sure as hell did not get rich today. The cost of bait was $14 per bushel, and I had used two totes — three bushels. I had not burned much fuel, but diesel was expensive these days at $1.43 per gallon. If I prorated both boat mortgage and insurance and divided the remainder into an hourly wage, I figured I could turn myself in to the Better Business Bureau for paying less than the legal minimum. It wasn't worth it to continue hauling traps. As soon as my father returned to the Island, we would start to "take up," or bring traps ashore for the winter. I climbed into the skiff and pulled the cord on the outboard motor.

In a season of good fishing and high prices, a hardworking lobsterman can make over $100,000. But a year like that happens rarely. For the most part, fishermen are lucky to cover expenses and make enough to keep them in food and clothing for the year. Seldom is there anything left over to put toward savings, especially in a year when major work is needed on a boat, which is just about every year.

I was figuring pounds of lobster and prices in my head as I motored toward the dinghy dock, but then began wondering if I had remembered to shut off the boat's batteries. I also wondered whether I had closed the valve on the thru-hull fitting for the deck hose. Should I go back and check? I probably had remembered both. It would only take a minute to be sure. I was tired and wanted to go home. I would come out in the morning and check the boat. It would be fine. What were the chances that the deck hose would fall off the fitting? It was double hose-clamped. I wanted to go home and call my mother. I would sleep better tonight if I went back to check the boat now. I was shaken from my internal argument as I realized that I was about to run into the stern of Payson's boat. I swerved to avoid putting my skiff's bow between the *v* and the *e* in *Perseverance*. If this was Divine intervention, I thought as I headed back to the *Mattie Belle*, God was not being subtle. I would mention it in tonight's letter.

MY TWO HUSBANDS

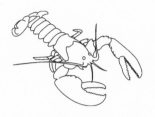

As it turned out, roasting red peppers in the toaster oven was not a good idea. The peppers themselves may have been fine, but it had become obvious with the blaring of two smoke alarms that toaster ovens are not self-cleaning. The blanket of crumbs that had lined the bottom of the oven had burst into flames and now looked like incinerated mouse droppings. It was too cold outside to tend the grill on my parents' deck, and I suspected the propane bottle on the side of the house was waiting for the most inopportune time to register "empty." Tonight would be the most inconvenient and therefore most likely time to run out of gas in mid-meal preparation, as I had planned to invite people to come for dinner.

In an attempt to silence the alarm, I stood

on a chair and fanned one of the noisy white hexagons with a *Bon Appétit* magazine opened to a recipe I thought I could handle, and wondered what I would eventually end up serving. Microwaving a pork roast was indeed as "ridiculous" as my mother had stated when I called her to collaborate (receive advice) on my dinner party menu.

A factor in this sudden domestic exercise of culinary art and adjunct cleaning spree was that tonight I would get to know my husbands. That's right, plural, two of them. We were not actually married yet, the three of us, but if Rachel Harris was right, I would have to marry both of the brothers from New Jersey who now visited the Island, as they were, she said, inseparable. Dinner would not be a double date, as I had no intention of including another single female. Available men only, the odds would be stacked in my favor. This would be fun.

First impressions were important, I thought, as I opened windows to clear the smoke from the kitchen. Tricking the men into believing me to be a good cook would be helpful. I stood gazing into the refrigerator, as if something may have appeared since the last time I looked. I searched my mother's cupboards for an inspirational ingredient. Capers, crystallized ginger, eleven different vinegars, seven mustards, polenta, wild rice, couscous, risotto, and pastas I did not recognize; I was certainly well stocked. If only I knew what to do with it all. Canned chipotles, ancho chili powder, four different cumins, two cilantros, a jar of Habaneros, tamarind concentrate; I wondered if Mom had been expecting a guest appearance by a chef like Bobby Flay. Hmmm . . . maybe I would boil some lobster. It's hard to screw up lobster, I thought.

Dating is tough on the Island. What do we do after dinner? I could take the guys out to jack deer with the one-million-candlepower spotlight that plugs into my truck's cigarette lighter. No, Annie Oakley never married, did she? Some men might not cotton to a woman armed with a loaded gun on the first date. Television was not an option. Television at my folks' place was more of a workout than it was entertaining. Yes we have TV. No we do not have reception. Thirty minutes of *Jeopardy!* would leave us with eyestrain and impaired hearing.

Maybe I should borrow a video from Ben, I thought, as I rummaged through my father's tools for a pipe wrench with which to shuffle propane bottles. Ben knew firsthand the scarcity of dates on the Island and would be helpful in choosing a film that might be a mood-setter. A movie would be appropriate for a first date, I thought. Would I sit between the two men on the couch? No movie, I concluded.

Making love with two men at the same time has never been one of my sexual fantasies. Hadn't I read somewhere that forty-year-old women are in their sexual prime? Probably written by a forty-year-old woman, I thought, and smiled as I pulled the proper wrench from under a pair of Dad's gloves. My sex life had been sporadic at best.

After twenty minutes of leaning on and beating upon the end of the pipe wrench and swearing at "the idiot" who had tightened the connection between regulator and bottle, I remembered that propane connections are reverse threaded. So much for "righty-tighty, lefty-loosy." "And this," I said through teeth gritted in pain as the 100-pound bottle came to rest on my toes, "is why God made men." I could definitely use two.

Maybe I was worrying too much about what to do on the date. I realized that I was far too busy to be spending time or energy wooing men. I still had four hundred traps in the water needing to be brought ashore, and it did not appear that my father would be returning to the Island soon. Aha! Why had I not thought of this earlier? I could track down the brothers from New Jersey right now and generously offer to show them a bit of the Maine lobster industry. I could do them the favor of allowing them to bring a load of traps in with me. We could spend all afternoon getting to know one another (working) and then have fresh lobster for dinner. If I really liked them I would attempt my mother's lobster casserole. If I could pull off the casserole, it would be worthy of at least two proposals, I surmised. I ran into the house to phone my mother for the recipe.

"Wow, you must be very fond of them," my mother remarked at the request. I did not bother sharing with my mother that since I hadn't yet met them I was merely anticipating the fondness; she is weird about giving out recipes. At this point I was taking for granted the men's acceptance of my yet-to-be-delivered invitation. Although my mother did not come right out and say it, she was not too excited about having me in her kitchen while she was away having cancer treatments. I scribbled as she recited. "I've never made this for less than six, so you'll have to cut it down. Cook and pick twelve lobsters, or sixteen if they are very soft. Linda, do not boil lobster in my large

Le Creuset." *Oh, you mean the one in the sink? The one that I have been soaking for two days to get the burned spaghetti out of? The one that I may have to take to Billings' to be sandblasted?* "In fact, do not use abrasives on any of my good pots and pans. I told you that before, right?" *Oh, you mean like scraping with a metal spatula? Too late.* Somewhere, through all of the marching orders and in the midst of many asides, I managed to pull a list of ingredients from my tight-jawed mother, but had to guess at amounts and temperatures. She was certainly less than forthcoming. The following is what I ended up with, but lacked the confidence to actually try by the time my mother was done with me:

8 tablespoons butter
8 tablespoons flour
4 cups light cream
A couple of egg yolks
1 handful of minced onion
1 generous splash Madeira
A little minced fresh parsley
Some salt
Some pepper
1 teaspoon celery seed
1 good dash cayenne pepper
12 cups lobster meat, sautéed

4 cups fresh bread crumbs
Parmesan cheese

Preheat oven to 400 degrees. Melt butter, and blend flour, cooking over low heat. Add cream and stir until thick. Remove from heat and whisk in egg yolks. Add onions, Madeira, parsley, and other seasonings. Add lobster meat that you have previously sautéed in butter. Pour into large casserole dish and sprinkle with bread crumbs and cheese. Place dollops of butter on top and bake uncovered until you think it's done (20 minutes at 400 degrees).

I would save this experiment for another night, I thought, as I hung up the phone. Tonight I would serve the foolproof and ever popular *boiled* lobster, the staple of my summer diet. I could boil a lobster with the best of 'em. My folks and I had eaten lobster once or twice a week all summer — boiled, sautéed, over pasta, in casseroles and omelets. Lobster is an inexpensive meal if you catch them yourself, and most fishermen enjoy the fruits of their labor while the price is low. I know of an Island fisherman who is allergic to lobster and feel sorry for him every time I indulge. Many things taste like chicken but nothing tastes like lobster.

The brothers were staying in a camp beside the pond. As I drove toward the East Side, I thought that I would need to stop at the store for a few things. This time of year it took a conscious effort not to allow the store to open and close again before I got there. The daily store hours were just barely plural, at two, and I feared that we would soon be faced with the "store hour." Residents complained that there was nothing in the store once the summer people departed, and many year-rounders had resorted to shopping for groceries on the mainland. It was a classic catch-22. No one shopped at the store because there was little to buy, and the store, because of lack of business, could not afford to stock the shelves. Apparently the store could not afford to pay an employee to hang around for more than a couple of hours each day either, as the store was now open only from one-thirty to three-thirty. I could, I thought, buzz the *Mattie Belle* over to Stonington for some groceries. But I liked to support the store.

As I drove by Theresa Cousins' house, I noticed another near catch-22. Two large, clear plastic garbage bags full of trash sat at the end of Theresa's driveway, where they had been since late September. Because of the severe drought there was a statewide ban on all burning. But the folks who had the contract with the Island to pick up trash are not responsible for taking "burnables." I had heard that Theresa left the bags as a way of making a statement. I was not quite sure what the statement was, but the overstuffed bags that drifted back and forth across the end of the driveway with the wind like giant tumbleweeds made it quite poignant.

I could amuse myself with these little games all winter, I thought. I could plan my days around trying to catch the store open, which was indeed challenging. And I could enjoy the calculated standoff involving the homeless bags of trash by counting the days that they remained before one party or the other gave in. Who says there's nothing to do out here? I was beginning to appreciate the satellite dishes that had been popping up in some year-round yards. What did they do out here in the winter before electricity, I wondered?

I wished that Rachel Harris had thought to install a telephone at her place, as I continued to drive toward the pond. Going to call on men whom I had not met was more difficult than placing a call to them would have been. I sensed cold feet at the end of my legs. Was I really going to barge in on the brothers who were vacationing at their friend's camp and invite them to work? I pulled into the gravel pit and placed the truck in park. What was I doing? The men might not even be home. They certainly would not be sitting around waiting for an unexpected dinner invitation from a strange woman. What if they did not want to haul traps? What if they did not want to come to dinner? What if they did not like the way I looked? As I checked myself out in the rear-view mirror, I wished I looked like someone else.

Once I had passed my last chance to bail out, I grew apprehensive again. I hoped the men would not be home. I could leave a note. Rachel's car was beside the road, parked in its usual spot. "Shit. They're home." Parking my truck, I took a deep breath. As I approached the door, I heard voices from the edge of the pond. Peeking around the corner of the camp, I saw two men with fishing poles. "Oh, thank God. They fish!" I ran back to my truck and grabbed the Orvis fly-fishing rod that had been rattling in the bed for months. Here was something I could do. What a relief. My gear was rigged for salt water, but who would know? I was absolutely delighted and suddenly confident in my approach.

I called a greeting as I walked toward them. The men were nice looking and friendly. Stripping line onto the ground at my feet and throwing the fly out to join theirs, I learned that they had been catching brook trout. They had fished every stream and pond between the Island and New Jersey, and were quite sick of trout, for they had eaten it three times a day since they had started their vacation trip two weeks ago. "Do you eat lobster?" I asked.

Within ten minutes, four casts, and two trout, I had both men in my truck, headed for the dock to catch dinner.

Dorothea Dodge

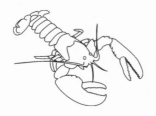

Through a very small service window in the very small post office on our very small Island can be seen a very small woman. The opening of the window through which all of the Island's postal business is transacted is so small, in fact, that only a glimpse can be had of a very small portion of Dorothea Dodge. It was only by chance one day that I happened upon our postmistress out walking her dog, and saw Dorothea in her entirety, confirming that she is indeed whole.

Dorothea, whom Islanders know as "Dotty," could appropriately shorten her name to the one syllable, since "Dot" accurately describes the amount I know of her. It is not that I am not interested or dislike Dotty, quite the opposite. But Dotty is the epitome of privacy, revealing absolutely nothing of herself in our casual postal dealings. More shy than aloof, Dotty has always been polite and friendly, but never initiates conversation. She runs a tight ship in the post office. Unless I am receiving a package or some article of mail requiring my signature, all I get from Dotty is "Hi," which I believe to be a bit more than some others may get, so I think Dotty likes me.

Dotty and her husband, Stanley, raised twin girls on the Island, Donna and Dianna. The girls and their husbands live across the road from the post office. Their houses cannot be seen from the road. Donna is the non-waving driver of the school bus. Dianna has goats and raises pigs. My friend and fellow fisherman Lincoln Tully is Dotty's grandson. Landon Dewitt, another grandson, goes to high school on the mainland.

Dotty lives alone in a tidy house behind the post office that she heats with wood. (I know this only from observation.) Because I have seen some of her artwork at a crafts fair, I know that Dotty paints. And this is the lump sum of what I know of the very small woman who has been an Island resident since before I was born, who is one of two score year-rounders, and whom I see six days a week through the very small window of the very small post office. Once when

Stanley was ill and eventually passed away, Dotty was absent from the post office for quite a stretch. I do not really understand what I missed of her in her absence, but I did miss her; and we were all greatly relieved when she resumed her post.

FIRST LIGHT

Before opening my eyes, I listened. The wind outside the bedroom window did not screech or howl through the trees, but was more than a rustle or whisper. Twenty knots out of the southwest, just as the man in the box had predicted. The last thing I do before going to bed each night is listen. And the first thing I do each morning upon waking is listen. At night the after-dinner rituals include the weather radio, or box. The computer-generated voice was a continuous cycle of climatalogical data: predictions, record highs and lows, rainfall amounts, Canadian observations, barometric pressures, and what I was most interested in: marine forecasts. Come morning I listen to hear if the man in the small, veneered, antennae-ed box has been correct in his predictions of wind velocity.

Now I heard the wind chime on my parents' deck. The steel against steel was a bit softer than the navigational bell buoy it had been manufactured to replicate. Maybe 25 knots, I thought, as I stretched and rubbed my eyes, preparing to open them.

I had not been sleeping well, and was distraught about it. Rarely having struggled with sleeplessness in the past, I took pride in my ability to sleep at will. The capability to doze on demand I considered a gift from God and perhaps the only natural ability I possessed. Now, the fact that I was having trouble sleeping was keeping me up at night. I worried about what might be on my mind. Sleep deprivation I have always been able to deal with. But lying awake when there is time to sleep I found frustrating.

How many hours had I lain awake last night, I wondered? The theater defined by the 3-foot-square skylight above my bed was host to stars that glinted like phosphorescence in the wake of a speeding boat. It had been a spectacular, moonless, star-stippled night. A galaxy of stellar candles burned holes through that black curtain of space, sparking imagination and illuminating possibilities. How many hours had it taken Orion to creep through my view? Night after night the archer hunted with loaded bow, making neither sound nor sudden motion. Orion was a great hunter, I thought. But the hunted must surely be craftier, as Orion never returns with prey. Greek mythology maintains that Orion pursues the Pleiades, the seven daughters of Atlas. He hunts for seven. I hunt for one. Perhaps Orion searches for something that will never be found. I wondered whether Orion had ever entertained the possibility that his quest might be futile. Maybe he knew but continued to make the rounds because it was what was expected. Or perhaps he had nothing else to do. I might have something in common with Orion, I thought. Although neither of the brothers from New Jersey had been interested in me, I had not lowered my sights. I was still looking, and still hopeful that I would have the opportunity to take a shot. And there was always that Vineyard Charter Boy I had been told about.

The New Jersey brothers had, in fact, just left for home and in all probability would not return to the Island until next fall. I had become accustomed to having them around, and although they had only been gone one day, I missed them. They had been so nice about helping me bring lobster traps ashore in exchange for dinner each night. We ate a lot of lobster. The days that were too windy to "take up" gear we spent hiking and exploring. The men clearly loved the Island, which rekindled my feelings for it. Like some child's toy that is neglected until a playmate finds it amusing, my home suddenly became something I was interested in. When the brothers announced that they must return to their home and work, I was both relieved and sad. Even though the time we had shared had not been romantic, it was nice having them around.

My father was here now. He had come to help take up gear, but typical of November, the wind blew relentlessly, keeping the *Mattie Belle* on her mooring. I was not sure how I felt about my father's arrival, relieved or sad. I had indeed spent far too much time alone this fall, so I was delighted to have Dad's company. His presence in the house made it warmer and more comfortable. He had come to help me, leaving my mother on the mainland to plod through her third cycle of chemotherapy.

Off-season proper was only days away, and I had a few traps yet to be taken up. I hadn't much of a plan for winter, which I found scary. Since I had been old enough to work, I had never been unemployed. Because I had always had another boat or season lined up prior to the end of the last one, I had never looked for work. I had never interviewed, filled out an application, or put a résumé together. If I did nothing all winter, the money I saved this season would be gone by spring, and I would have to start from scratch. I was not enjoying what I anticipated unemployment would feel like and hoped some interesting opportunity would fall into my lap, saving me from learning how to get a job. There were no winter opportunities for me on the Island.

At the end of the season, lobsters were again proving fewer than scarce, and the price was dreadfully low at just $2 per pound. The Association's pound was full — 40,000 pounds of lobster. And there was the pervasive fear that the price would not recover its usual level due to the floundering economy.

I had gotten over my anger with the members of the Association who had backed out of the effort to obtain an exclusive fishing zone and wimped out of the gear war. Most of the men preferred the status quo and prosperity to what promised to be an uphill, expensive, and possibly dangerous battle. I was a bit sad, though, and vowed to try next year to help resurrect Jack MacDonald's vision for the preservation of our precious community. "Beach glass" Jack would say, or "ocean effect . . . moderates things . . . no extremes . . . no rough edges." Jack knew.

I felt guilty about enjoying having my father around. It took a conscious effort on my behalf not to sound too happy when speaking to my mother during our phone conversations. Mom was having trouble releasing her tightfisted grip on all that transpired within and around the realm of the family. She somehow mustered the energy to call at least once a day to give me my "marching orders" and once each night to check up on my progress executing them.

A true matriarch in the most traditional sense, my mother now lacked the strength to do *everything* herself. She had conceded the Thanksgiving grocery shopping duties, but did supply me with her list. She demanded that all ingredients be delivered to her house on the Tuesday prior to her arrival for the holiday. She planned to cook all day on Wednesday and requested my presence. If she was not physically up to preparing our favorite recipes, she would, she said, lie on the couch and boss me through them. I agreed and was thrilled that she would be well enough to at least be bossy.

I smelled coffee and decided to delay opening my eyes for just a few more moments to enjoy the aroma. Since climbing into bed, the past eight hours had been odorless. All smells I now associate with the light of day. Darkness has no smell. Oh, it has not always been this way for me. In years of swordfishing, the hours between dusk and dawn had been permeated with the smell of thawing squid and salt air warmed by the Gulf Stream. Nights ashore between fishing trips were often spent relishing the scents of celebration. Barroom stenches of cigarette smoke and booze were never offensive until the light of day. How long had it been, I wondered, since I had inhaled the mustiness of company in bed? Sleep was now odorless. I feared that I might fall into the footsteps of Alfreid Thompson, the old maid who raised sheep in Head Harbor back in the early 1900s. When haggling over a price with a gentleman from away, they reached a point from which neither party would budge. When the gentleman suggested they "split the difference," Alfreid exclaimed, "I'll have you know that the difference hasn't been split in fifteen years . . . not since Captain Henry died!" I certainly did not want to end up like that. I have to get up, I thought, and forced my eyes open.

There was not yet enough light to cause me to squint. The predawn world of dappled grays is soothing and soft on just-opened eyes. All I could see, both inside and out, looked dusky and leaden. To open one's eyes after eight hours of blackness to a day in full bloom can be offensive. Vivid, clear, intense color all at once is hard to take. Much better, I thought, to watch the day slowly ripen from blanched and anemic to bright.

Dawn is a kind of spring for each day, even in the dead of winter. New beginnings, renewed hope: All of the worn-out metaphors and analogies for spring are applied to sunrise by "morning people." I love mornings, especially in the fall. This time of year, deer and ferns are both the color of a russet potato.

From my perspective, still on my back in bed, I could see the silhouettes of trees dancing in the wind. Entire trees gyrating. The larger, fuller branches appeared to be bumping and grinding with some enthusiasm, the breeze and first light aphrodisiacs. Eventually I rolled out of bed and onto my feet. Here on the West Side of the Island, we see only the shadows of sunrise. West Siders live in Plato's cave every morning until the sun rises above the mountainous ridge. I knew the sun was coming up without actually seeing yolk rise from saltwater white. It was something I took for granted.

I threw some clothes on and skipped down the spiral staircase to join my father, who sat sipping coffee and staring at the ocean. "Hi, Lin."

"Good morning, Dad." I poured a cup of coffee, added excessive cream, and sat facing the water. The wind was out of the south and had teased up a 3-foot sea. Battalions of waves paraded by, heading north. I could not see where they stormed the beach, but imagined it might be just above the pebbly spit of Money Point. Some of the officers on horseback nodded shocks of white hair while masses of lower-rank sailors kept eyes forward and sternly marched in the most rehearsed fashion to the wind that gusted and relaxed with the beat of some patriotic song. I could not hear the tune but felt the rhythm of booted feet hitting ocean floor. The trees lining the shore waved like spectators, some with children on their shoulders. Some of the old guard even saluted.

"Not much of a day to take up traps," my father remarked. There was so much to be done before Thanksgiving. We would not be bored. It was good to have days when the wind kept us ashore to take care of other things. We expected thirty guests for turkey dinner. Everyone, friends and family alike, would contribute something to the feast.

The day promised to be quite festive, Thanksgiving and reunion all in one. It would be the first time my mother would be home since the start of her cancer treatments. We all certainly had plenty to be thankful for.

"I'll grab that collapsible crib and take it to Aunt Gracie's. Chuck, Jen, and the boys will stay with me there." My eight-month-old nephew, Addison James Greenlaw, would sleep in the crib that I would set up in the same room in which his parents would stay. His three-year-old brother, Aubrey, would no doubt want to sleep with his "Aunt Linnie" in the other bedroom. I was anxious to see my brother and sisters in spite of the fact that they were using all of the good old family names for their children, leaving me with those I considered less than attractive. Avis? Eugene?

Six acres of land had changed hands from my parents' to mine and a house lot would soon be cleared. Part of my dream was taking shape. I had always envisioned moving into my first home as a complete family unit. I had almost come to terms with the fact that I would probably move in alone.

With no baseball glove left in the middle of the floor and no milk spilled on the counter, I wondered if my house-to-be would ever feel like a home. I had never imagined a home without children. People are sometimes taken aback by my openness in desiring children and family. I am not embarrassed or ashamed about wanting to re-create what I had growing up.

Dad had quite a list of stuff, lumber and hardware mostly, to be picked up on the mainland. I volunteered to run to Stonington for supplies that included the items on Mom's grocery list while Dad stayed and worked around the house. He dropped me at the dock and agreed to meet me in three hours to load things from boat to truck to house. As I left the thoroughfare, I was able to smile. The ocean had always had the ability to swallow my troubles. For the next 7 miles and 30 minutes I would not think of cancer, unemployment, loneliness, children, or lobsters.

As I turned the *Mattie Belle* north, between Merchant's and Hardwood islands, the southwest wind was kinder. Wind on the stern, or "running before it," is a contradiction of sorts. With any amount of wind, putting it on your stern is the most comfortable direction in which to travel, but the helmsman concedes some control of the vessel. Gently lifted and carried, I knew the ride back into it would be harder going. The ease of leaving the Island was magnified when I looked over my shoulder to see that its mountainous ridge had vanished in the poor visibility. The ocean continued to nudge me away from the Island. Like a mirage, I thought. Now you see it, now you don't. Gently lifted and carried, lifted and carried by the sea, I was soon in Stonington.

The trip back was indeed harder going, as the conditions had deteriorated. The wind threw heavy spray up and over the bow, and salt water poured from the back of the overhead. A Newfoundland powerwash, I thought. Wind and sea tried to drive me back to Stonington, but the *Mattie Belle* was headed to Isle Au Haut. Roughly pounded and shoved by the sea, I made my way slowly into the lee of Kimball Island. The top of the Island reappeared, exactly as it had been when it had vanished.

Dad was waiting with the truck backed onto the dock. We worked together silently loading groceries, lumber, and hardware from the *Mattie Belle* into the back of the truck. We went from dock to mooring, and from mooring to dock, exchanging the *Mattie Belle* for the skiff. I climbed behind the wheel of the truck while Dad slid in on the passenger side, and off we went toward home.

Epilogue

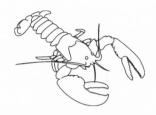

As of this writing, I can't report that I am married, pregnant, catching lots of lobsters, and living happily ever after. . . . My life is now one big loose end. The only conclusion I have been able to draw from my new life ashore is that I have achieved normalcy only in the sense of being in the same state of confusion that characterized my life at sea.

My mother has, as of today, finished her chemotherapy. She has lost far too much weight and, of course, all of her hair, both of which promise to return with a vengeance. Up until the last few months, I never would have described my mother as being particularly brave or tough, but now those are the first two adjectives that come to mind. (She has, I think, a crush on her young oncologist, Dr. Keating.) Seven weeks of radiation begin soon, and we all look forward to next summer's celebration of cancer survivors. My mother assures me that she will be leading the parade, and of that there is no doubt.

My father and I had all the traps out of the water, and the *Mattie Belle* high and dry at Billings' in the first week of December. We will not give lobstering much thought until May, when the buoys get a new paint job. This season fizzled, or I should say that we did.

The off-season Island is quieter than usual this year. There are presently forty people in residence. I see the same five every day at the store and wonder where the other thirty-five are hiding. A special town meeting drew fourteen adults, and we spent more time voting in a moderator and debating compensation than we did on the articles of said meeting. The school's Christmas program was preceded by a potluck dinner, and most of the forty were present. The seven students performed and made quite an entertaining evening. Unless someone starts having babies out here soon, the Christmas program will become more challenging as the number of participants shrinks with graduations.

Sadly, Jack MacDonald, who had fought so hard to protect our fishing rights and in doing so had become my hero, passed away, leaving quite a void on the Island. Gordon Chapin, our oldest fisherman, also died. Gordon and Jack are missed by all Islanders, seasonal and permanent, and by the mainland fishermen among whom they fished.

Dad and I ended up repairing my shop's roof, as the Island Boys remained elusive until the last nail was driven. The pair has offered to help with the construction of my house, but now I am doing my best to be the elusive one. I am building a year-round home, but am undecided about how much of the year I will stay. My best friend Alden is having a new boat built, and that has me thinking of going offshore again. I would like nothing more than to go swordfishing again. It would be nice to go just because I want to and not because I have to. As proud as I am to say "I'm an Islander," nothing makes me prouder than to say "I'm a fisherman." And that is not apt to change.

ACKNOWLEDGMENTS

Anticipated thanks to my family and friends for not banishing me from the Island upon publication of this book. You can now relax in conversation and stop warning one another in my presence to "Be careful of what you say. It might go in the book." Thanks, too, for your continued interest and support when the writing took longer than any of us imagined possible.

Thanks and appreciation to all who helped with information, formal and informal. To name a few: Wayne Barter, Lucinda Russell, Duane Pierson, and Murray Gray. Special thanks to Ben MacDonald for his friendship and illustrations.

Thanks to my literary agent, Stuart Krichevsky, who has once again proven himself a man of great patience and wisdom.

Thanks in advance to my best friend Alden Leeman, to Charter Boy, and to Captains George and Tommy for not filing lawsuits.

Thanks, respect, admiration, and adoration go to my editor and friend, Will Schwalbe, the rudder of my writing process.

Thanks and love to Mom and Dad for everything.

MAP OF ISLE AU HAUT

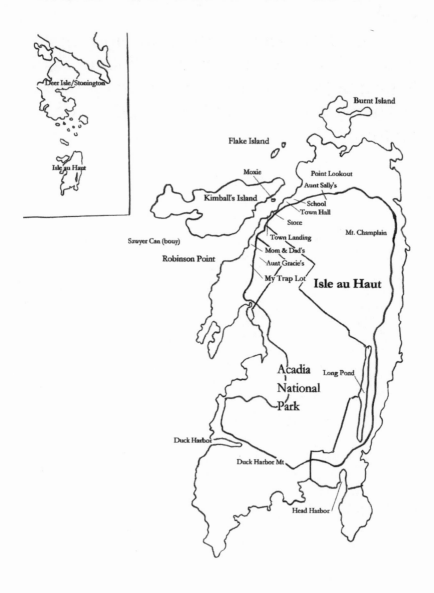

Deer Isle/Stonington

Isle au Haut

Burnt Island

Flake Island

Moxie

Kimball's Island

Point Lookout

Aunt Sally's

School

Town Hall

Store

Sawyer Can (bouy)

Town Landing

Mom & Dad's

Robinson Point

Aunt Gracie's

My Trap Lot

Mt. Champlain

Isle au Haut

Acadia

National

Park

Long Pond

Duck Harbor

Duck Harbor Mt

Head Harbor

DATE DUE